THE REDUCTION OF PHYSICS TO TIME

Spacetime and Matter constructed from the Arrows of Time

By

Carey Ralph Carlson

THE REDUCTION OF PHYSICS TO TIME

ISBN: 979-8-90252-301-7 (Paperback)
ISBN: 979-8-90252-300-0 (eBook)

Printed in the United States of America

Contents

INTRODUCTION

Overview: The Reduction of Physics to Time

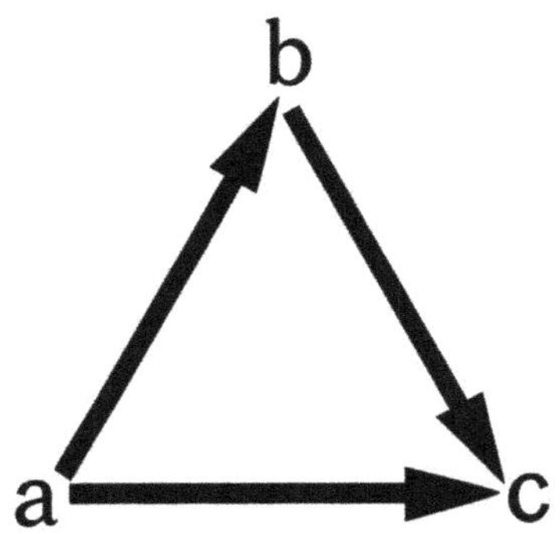

I was walking down the basement stairs one day in 2003 when it occurred to me that time can make frequency ratios. In my mind there was a directed graph of three arrows forming a triangle.

There are two paths from "a" to "c". One path takes two steps (a to b to c), while the other path takes one step (a to c). The two paths begin at the same time and end at the same time. Thus, the two paths transpire in the same amount of time. One path takes twice the number of steps as the other path to traverse the common time span. The stepping rate of one path is twice that of the other path. That is a frequency ratio of 2:1. Time can make frequency ratios. It's a momentous discovery.

Before the limiting velocity was discovered, time was thought to be one-dimensional. One-dimensional time could not make frequency ratios. But we see in the example above that the branching and converging pathways of causal set time can indeed form frequency ratios. The frequency ratios, in conjunction with Planck's E = hf, serve to define energy ratios. The step of time is the unit of the frequency ratios and the quantum of the energy ratios. We have energy and its quantum defined in terms of time alone. Arrows of time graphs then serve as quantum schematics, with each arrow representing a quantum. That is the incentive for the constructions to follow.

The reduction to time is a reduction to standalone causal sets. Everything in physics is part of the causal set of the universe, which can be imagined as "the arrow diagram of the universe." A causal set is analyzable into "elements" and "causal links." Each causal link connects one element to another element, establishing the time order of a predecessor/successor pair of elements. You might say that the universe is generated by a causal successor pairing relation. The causal links are instances of that relation. The universe is a concatenation of causal links. In the arrow graphs, "elements" are the junctions where arrows meet. The junctions are the locations of the causal web. Thus, the proposed reduction is to the two causally primitive parts of the causal web, plus the structural formations comprised of those parts. No further parameters, numbers, or entities are needed. All the number and structure of theoretical physics are inherent in the bare causal set skeleton of the universe.

I said above that the reduction is to causally primitive parts. The "elements" have an interpretation in Whitehead as momentary "occasions of experience." The sentient occasions have phenomenal ingredients, which means they are not ontologically primitive. But occasions are causally primitive. They function as the primitive causes and effects of the world. The universe of sentient occasions accommodates human minds that can make phenomenal observations and concoct theories. The restriction to formally primitive parts, without provision for phenomenology, stands on its own merits as a formal theory of sequential cause-and-effect. That explains the demarcation between mathematical physics and its underlying metaphysics. The latter ventures an interpretation of the causal primitives at the basis of the former.

The simplest and most symmetrical graphs produce the most fundamental entities and features of physics, the 4-D time manifold, the electron and its cloud formations, the FSC, Bohr's formula, QM, and the proton and neutron with their mass ratios to that of the electron. That cannot be an accident. It is evidence for the reduction.

The constructions are relativistic by default due to the exclusion of primitive spatial relations from causal set theory. The reduction to time entails a limiting velocity because any change of space-like distance is a purely structural consequence of the stepping action of time (space cannot "outrun" time). A discrete time dilation, from one atomic cloud to the next, is employed in the graphical account of Bohr's formula. That should provide the basis for discrete GR (see the Appendix). The manifold, like everything in the theory, is made of quanta, which suggests the source of dark energy. With the ratio-based metric, the "Planck scale" loses its importance. Any quantum can serve as the reference for the frequency/wavelength ratios of all the other quanta. Symmetry is the only organizing principle for standalone causal sets. Symmetry governs the formation of particles and produces the laws and forces, effectively assuming their role.

The theory is fully discrete. The wave-particle duality is resolved by the discrete construction of the particle sequences. There are no continuous waveforms that require collapse, and no spatial states to which waveforms could collapse even if they had existed. A 4-D time lattice structure is used to graph electron clouds and neutrino modes of propagation, which together comprise the 4-D manifold. A 6-D time lattice structure is used to model the nucleons. The integration of the hydrogen nucleus with its electron cloud affords the calculation of the mass ratio of the proton with respect to the electron.

The remainder of this book brings together selected papers from the past 20 years in detailed support of the reduction to time:

“A Theory of Everything for Physics, Second Edition.” (booklet, originally published in 2005)

“Causal Set Theory and the Origin of Mass-ratio.” (paper from 2010, adds the nuclear structure)

“A Frequency Ratio Account of Temporal Atomism.” (originally published in Process Studies, 2020)

“The Primordial Nature of Today’s Fundamental Particles” (2022)

“The Role of the Causal Set Proton in the Formation of 3-D Matter” (2023)

“Discrete Time Dilation from Causal Sets and Its Role in Discrete GR” (2024)

Physics / $8.95

The cover diagram depicts the repetitive structure of a 4-dimensional lattice, based on a stepwise progression that may be interpreted as "temporal succession." The arrows therefore map the routing of time flow from the two earliest moments at the bottom of the diagram to the two latest moments at the top. Major insight comes when we put "physical space" out of our minds altogether and think in terms of "4-dimensional time."

Each arrow represents a single step in temporal succession and a single quantum of energy. The number of nodes in the diagram determines a calculable number of alternative sequences, yielding variants that differ in total number of arrows. The energy inherent in a lattice may therefore fluctuate from one location to another.

The "scale" represented by the diagram is indeterminate, awaiting an assignment of measurable duration to the diagram as a whole. At low frequency it can span an astronomical scale in the theory of gravity. At high frequency it can span an atomic scale in the theory of the atom. It should be useful wherever four-dimensionality is found. No matter what the scale, nodes are available as termini for sequences of disparate frequency, so that all sequences may connect in one consistent history of temporal succession.

CAREY R. CARLSON is a writer based in Minneapolis. He studied the philosophy of science under Grover Maxwell.

ISBN 0-929636-40-6
50895
9 780929 636405

Syren Book Company
MINNEAPOLIS, MINNESOTA

A Theory of Everything for Physics

SECOND EDITION

With a Topological Derivation of Atomic Spectral Data

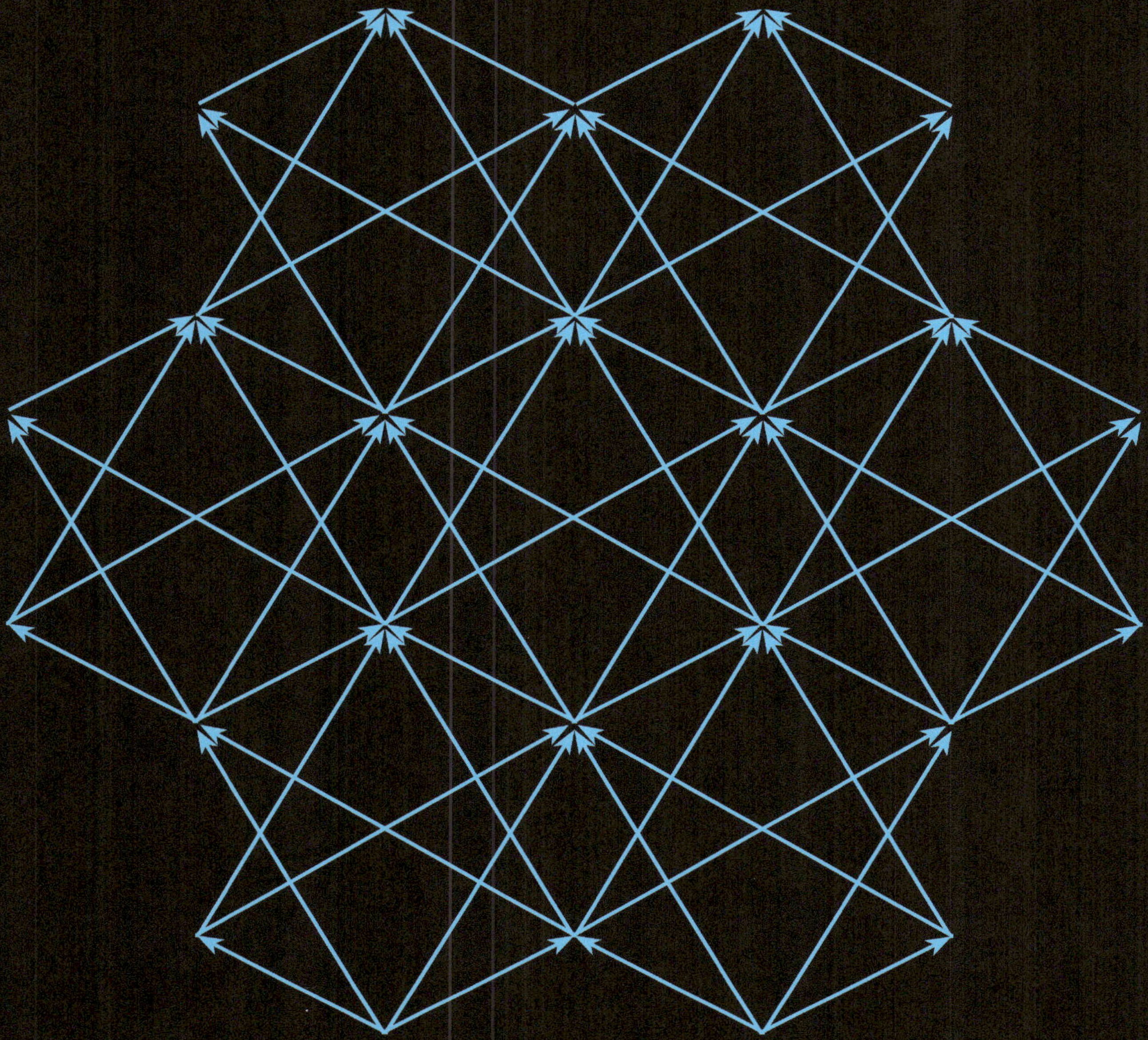

A THEORY OF EVERYTHING FOR PHYSICS

"Space is made of time." At first encounter, that statement appears to be nonsense. Yet it summarizes the doctrine of space-time that Russell and Whitehead have advanced in their writings. I wanted to give formal precision to that doctrine because it was central to the thesis of the book I was writing. I refer you to The *Mind-Body Problem and Its Solution* for the full supporting argument. The Theory of Everything that coalesced in that book is extracted here and presented with a minimum of interpretation.

I made the choice to treat the structure of time (or cause-and-effect) as discrete rather than continuous and capable of infinite divisibility. That choice was initially made for ease of explanation, but it proves to be the premise of quantum theory.

The theory concerns the time ordering of a class of primitive individuals. By ruling out infinite divisibility, the discrete structure of any time sequence is fully specified by the next-to-next succession of its individuals.

We shall use the terms "successor" and "predecessor" to mean "immediate successor" and "immediate predecessor such that there is no event between a predecessor-successor pair (no third individual that succeeds the predecessor and precedes the successor). Ordered-pair notation may be used to express the primitive facts of the theory, so that "(a, b)" states that "b is a successor of a." Event "a" may have more than one successor, such that "(a, b)" and "(a, c)" are both true. This is a "fork" in time. We must provide for convergence as well. For example, “(d, f)” and “(e, f)” may both be true, in which case “f” has two distinct predecessors. The forking and convergence of temporal sequences will account for the branched-out structure of space-time. We shall not rule out any finite number of predecessors or successors.

The theory may be conceived as the logical conjunction of its primitive facts, which are all of the form “(x, y).” The ordered pair is a relation of two terms. We stipulate that this “temporal successor” relation be asymmetrical, such that “(x, y)” and “(y, x)” cannot both be true; and irreflexive, such that “(x, x)” cannot be true.

We next define “earlier than” and “later than” in terms of the successor relation.

If (a, b), then a is “earlier than” b.

If a is earlier than b, and b is earlier than c, then a is earlier than c (transitivity).

If a is earlier than b, then b is not earlier than a (asymmetry). Similar statements define “later than.”

That should ensure that no individual is earlier or later than itself, and that no time sequence loops back on itself (there is no “going backward in time”).

We shall refer to the foregoing specifications as “the basis” of this theory. The ultimate parts of the physical world, its fundamental entities, have been fully specified, and no others will be introduced. The only fundamental entities are physically featureless individuals and the relation of temporal succession that orders them. That relation, together with an unlimited supply of individuals, offers endless structural possibilities that may characterize a history of temporal succession. We shall find the “meat” of scientific theory among these structural possibilities.

Now we introduce another notation that is better suited for further development. Examples of time sequences can be specified by using arrow diagrams.

A single arrow depicts a primitive fact of the theory, so that "(a, b)" is expressed: a ⟶ b

The temporal sequence "(a, b) and (b, c)" looks like this:

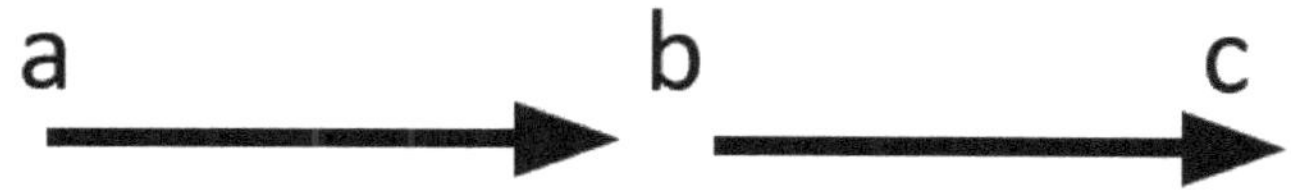

Figure 1

Arrows concatenate to one another only at their heads and tails, such that each junction of arrows represents a single individual. We can recover the ordered-pair notation for any arrow diagram by labeling its individuals uniquely and then translating each arrow, one by one, to an ordered pair of the two individuals at its endpoints. We have no further use for the labeling of individuals, so we can stop doing it. The individuals are just "dummies" in the theory, differentiated solely by their relative positions in some temporal sequence. Just think of the individuals as "moments in time," and we will say no more about them. The axioms of asymmetry and transitivity dictate a rule in the construction of valid arrow diagrams: no directed path of arrows may return to its starting point.

Here is the example of "forking" given earlier as "(a, b) and (a, c)."

Figure 2

And here is convergence, given earlier as "(d, f) and (e, f)."

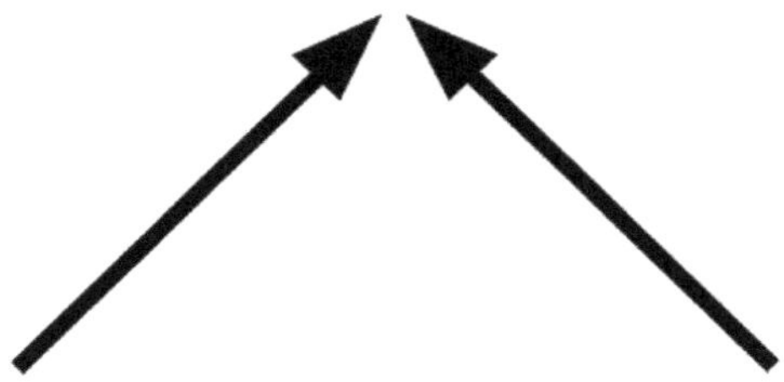

Figure 3

Next, we show forking and convergence together in one connected sequence:

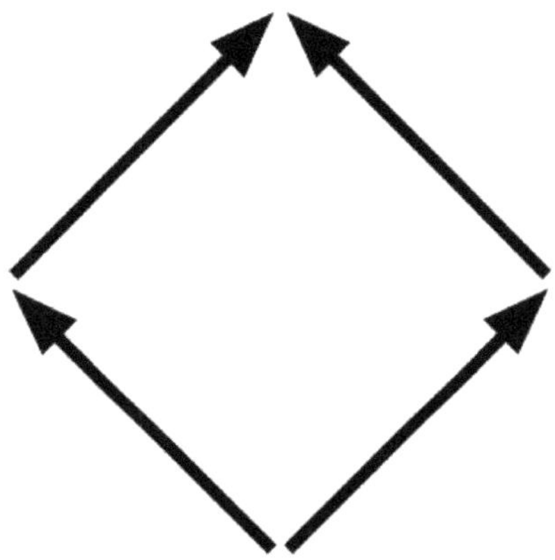

Figure 4

In Figure 4, four individuals make four succession pairings. The next diagram also involves four individuals, but they make five succession pairings.

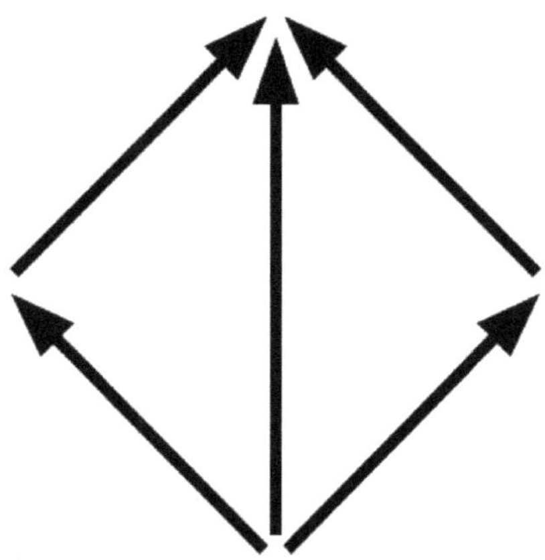

Figure 5

In order to specify the "hollow" structure of Figure 4, by which it differs from Figure 5, we need to list the four ordered pairs that define Figure 4, and we must also indicate that the list is complete. The same applies to arrow notation. We must be given to understand when a diagram is complete, that is, when all its arrows have been drawn.

In an important sense, all diagrams are incomplete. The principle of "one causally connected universe" is expressed here as "one temporally connected universe." Under this principle, the only truly complete diagram would be a diagram of the whole universe, a diagram that maps out the complete sequence of contingent events in full detail. Such detailed knowledge is an omniscience not available to us. We need to isolate small pieces of the world and focus on the most primitive formal possibilities of succession, by reference to the most rudimentary arrow diagrams. One tactic is to explicitly interpret a given arrow diagram as a "possible universe," complete in itself.

One way to organize the possibilities of time order is to start with a universe of one individual (blank diagram), then move on to a universe of two individuals (one arrow), and so on.

There are four possibilities for a universe of exactly three individuals, and three of these are shown in Figures 1, 2, and 3. The fourth is a triangle. In a universe of four individuals, two possibilities are shown in Figures 4 and 5, and there are several others to consider. In this theory, all things physical are to be identified as structural features of temporal succession, so we must suppose that the hierarchy of possibilities that we have begun to diagram, starting with the simplest, represents the "fundamental particles" of physics, while a single arrow depicting a discrete temporal transition represents the "quantum of energy." Not all quanta have the same energy, and we will need to develop a definition of temporal frequency before we arrive at a structural definition of measurable energy. We shall resume consideration of the fundamental particles later.

In Figure 6 we have an arrow diagram that resembles a conventional space-time diagram in a physics textbook, except that this one depicts discrete time transitions.

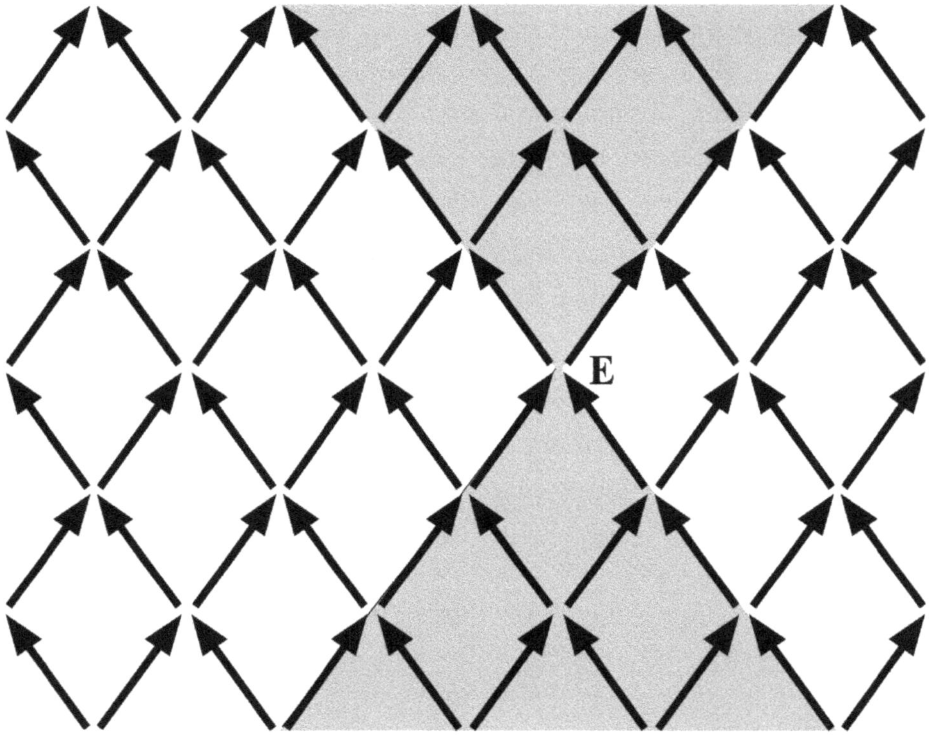

Figure 6

Event "E," an individual, is labeled, and its "causal past" and "causal future" are shaded to isolate a "light cone" with E at the vertex. If we imagine the diamond pattern extended uniformly in all directions, then every individual in this universe has exactly two predecessors and two successors. The piece of universe we're looking at is zoomed in to the quantum level. We can see the manner in which Special Relativity and Quantum Theory coexist in this universe. Quantum Theory arises from the premise of next-to-next succession, and Special Relativity arises from iterated forking and convergence. It is also clear, I think, that whatever *spatial order* is definable in Figure 6, it must be culled from the fully specified *space-time order* of Figure 6. The latter is specified in the formal basis as a logical conjunction of primitive statements pertaining strictly to time order. I offer that as a more carefully phrased version of my opening remark, that "space is made of time." All statements regarding spatial order are reducible to statements that pertain strictly to time order.

We can define another possible universe based upon Figure 6, identical except for the absence of individual "E." The four arrows that pinpoint E in the diagram, the four succession pairs to which E belongs, must disappear along with E. This creates a singularity in the newly defined universe, a "hole" in the uniformity, which is suddenly the most interesting feature of this universe. Because our own universe is "lumpy" with regard to the distribution of energy in space-time, we can base various diagram constructions upon Figure 6, and by judicious deletion of arrows and individuals, leave a connected skein of arrow patterns that better mimics the non-uniformity of our universe, as for example, Figure 7.

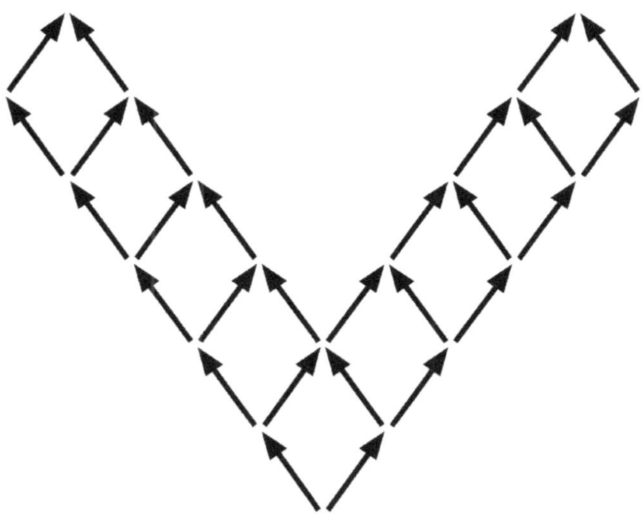

Figure 7

In Figure 7, we see something that might represent two particles with increasing space-like separation. In our theory, it is a circumlocution to say that a particle *has* a history. Conceived as a time-ordered sequence of moments, a particle *is* a history, and nothing in the theory can be said to "have" a history without risking confusion. The progressive separation of "particles" is subject to a limiting velocity because space-like separation is only achieved by time-like propagation. Any universe constructed according to our formal basis will have a velocity limit. The presence of a limiting velocity appears as an arbitrary complication to Newton's system of the world, but working from the formal basis that we have adopted, a velocity limit exists necessarily. This is trivially so because "time cannot outstep itself." We have no need for an axiom regarding the limiting velocity of the speed of light.

In regard to the "dimensionality" of space-time, our basis makes allowance for any finite number of predecessors and/or successors to an arbitrarily chosen event. A four-dimensional space-time can be constructed by concatenation of the structural element shown in Figure 8.

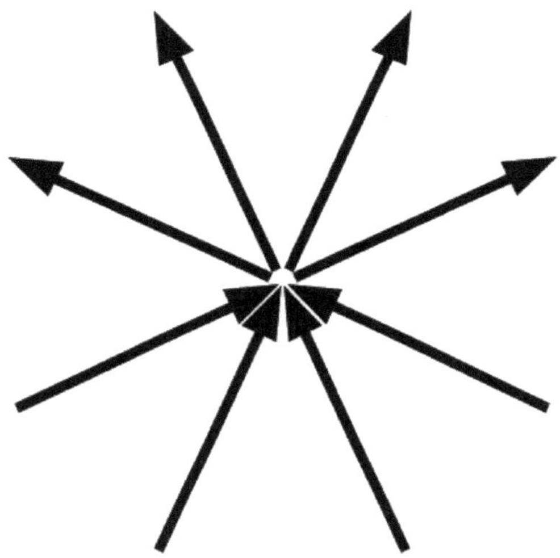

Figure 8

We could define a uniform latticework analogous to Figure 6, such that each individual has exactly four predecessors and four successors, as in Figure 8. From that basis, we could perform deletions to better model the nonuniformities of our own universe, in the manner that Figure 7 was obtained from Figure 6, but now set within a four-dimensional frame of reference.

Now let us attend to the definition of "frequency," and more specifically, the *relative* frequency of two causal chains that diverge from a single individual and reconverge at a single individual. An example is depicted in Figure 9.

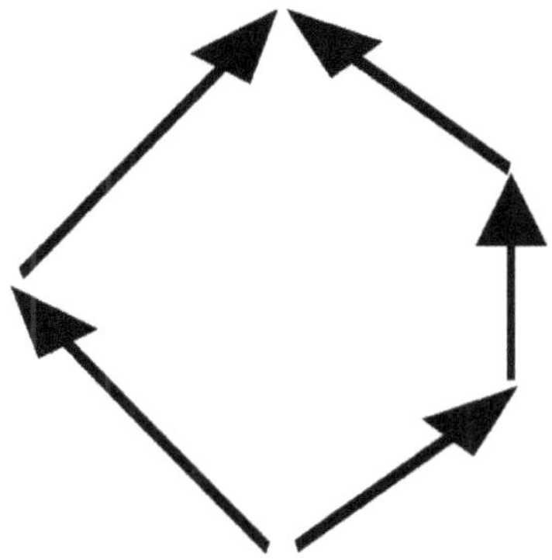

Figure 9

Figure 9 shows five individuals in five pairings. One time-like path from the earliest event to the latest involves two quantum transitions, while the other path involves three. *For two paths with common endpoints*, let us define the "relative frequency of one path with respect to the other" as the ratio between the arrow count of the first path and the arrow count of the second. In Figure 9 then, the path on the left has a relative frequency of 2:3 with respect to the path on the right. "Energy" will be assigned to the two paths in proportion to their relative frequency. The energy assigned to a path can be distributed evenly among its quantum transitions without doing any harm, as long as we keep in mind that no energy is definable in the first place without the condition of common endpoints obtaining for two distinct time-like paths.

Likewise, the "duration" of the bounded region determined by the first and last event of Figure 9 may be distributed evenly among the constituents of the left path, and also among the constituents of the right-hand path, without doing any harm. Because of the finite divisibility of any bounded region, a finite iteration of ratio calculations based on whole numbers will always yield a non-zero minimum value among the relative durations assigned to individual arrows. This is why I refer to an individual as "a moment," and not to anything as "an instant."

Notice that in the previous paragraph we are not introducing any new primitives into the drawing but rather we are "pulling out" features that will define "frequency" and "energy." That is why I refer to such derivations as "purely structural definitions." With regard to the metric by which we shall obtain measurable time, space, and energy, this too will be "pulled out" of our basis without introducing any new primitive.

Whitehead long ago gave an account of spatial order in terms of time order. His account goes something like this. Define temporal "betweenness" in terms of before and after, such that an individual is "between" two others in time if it is before one and after the other. Then two individuals are defined to be "contiguous in time" if one is before the other and no third individual is between them. He then defines "spatial betweenness" for three individuals that are mutually contemporary, by reference to the respective *pasts* of the three individuals. As in Figure 6, the "past" of an individual is the set of all individuals that are earlier than the given individual. For three contemporaries A, B, and C, if the past of B includes as a proper subset the intersection of the pasts of A and C, then B is defined to be "spatially between" A and C, and otherwise not. Then two contemporaries are "spatially contiguous" when there is no event that is spatially between them. From this point, a "contiguous region" of space-time is defined in terms of temporal contiguity and spatial contiguity. The intersection of contiguous regions can then be used to define "convexity." The intersection of convex regions is then sufficient to define points, lines, and surfaces. I am going to follow a different path that is more tenuous, and I defer to Whitehead for a more trustworthy derivation of space-time geometry from temporal succession.

It seems to me that any two time-like related individuals (one earlier than the other) determine a "bounded space-time region." They do not in general determine "a line." If Figure 9 is a complete universe of five individuals, then that universe consists of all the time-like paths that originate at the earliest individual and terminate at the latest. On the other hand, Figure 9 may be interpreted as a proper portion of some universe. In that case, the logical union of all the time-like paths bounded by the earliest and latest events of Figure 9 serves to isolate for discussion a determinate finite portion of its wider universe. I don't know whether such a bounded portion of space-time can serve as a "region" for the derivation of geometry in the manner of Whitehead, but I believe such "boundedness" to be important nevertheless in regard to the practical isolation of causal factors in a physical experiment. We have already paid respect to the principle of a "causally connected universe." Let us pay further respect to the more restrictive principle of a "causally closed universe," but only in a piecewise fashion. Let us refer to Figure 9 as a "closed" diagram. Any diagram that is bounded in time by a single earliest event and a single latest event will be "closed." All other diagrams are "open." Among the figures we have produced so far, 1, 4, 5, and 9 are closed. The rest are open.

In a closed diagram, we see the whole past and the whole future. Any two contemporaries are assured of both some common past and some common future. Their spacelike relations are doubly determined, once in terms of the individuals that are earlier than either of them, and once in terms of the individuals that are later than either of them.

Our definition of "energy" is inspired by Planck's constant, not by the decimal value but by the fact that the value is a constant. This fact suggests that temporal frequency alone will be sufficient to define energy. The relative temporal frequencies in a closed diagram are fully rationed to one another, but there is no suggestion of any "absolute frequency." (Our definition of time is relational, and that will infuse our definitions of frequency and energy with a similar "relativity," which I think is a good thing.)

To obtain what we think of as "absolute frequency" from the relative frequencies in our own neighborhood of space-time, we just look to the method adopted by science. We fasten upon some trusted repetitious causal chain in the real world as a useful standard, such as a laser or atomic clock in the Bureau of Standards, and reference all other frequencies to the adopted standard. The de facto standard of the "second" is a practical unit for the measure of duration, as good a unit as any. Event count per second is used in conjunction with Planck's constant to calculate the energy value of any repetitive sequence that coexists with the Bureau of Standards in a bounded region of space-time. Any bounded region will have a finite amount of energy, which can be apportioned out to each quantum transition according to the inverse of its duration, which is its "absolute frequency" in cycles per second.

An arrow is a "step in time," and it also doubles as the only "step in space" that we shall recognize. Once we've proposed to interpret some arrow path in a closed diagram as a representation of some specific trusted empirical repetition, we have a unit by which to measure its time intervals, and we have the same unit by which to measure its spatial intervals. Time-like intervals and spacelike intervals are well distinguished in the diagrams by whether or not they involve contemporaries. It takes two or more time-like paths to determine a spatial interval. I believe that these facts underlie the succinct power of "the square root of minus one" in defining the metric of Special Relativity. It is only when we combine two divergent time sequences that any spacelike separation is formed. If we have a time measure for such a pair of divergent time sequences, then a reasonable definition of measure for the spacelike separation that is formed is a normalized product of the two component time values involved.

If we attach the imaginary "i" to each component time value before we calculate the normalized product, then we get a real number for a spacelike interval only when two-time sequences are invoked to define the spacelike separation being calculated. Thus the derivation of a metric for space-time from our formal basis makes time-like and spacelike intervals distinct, yet commensurable in the precise manner required by Special Relativity.

We now have available, if all my assumptions are correct, a defined measure of energy in any bounded region, as well as a defined measure of volume for that region. The energy divided by the volume defines the "energy density" of a given region. This, I believe, is what Einstein correlates to the "curvature of space" to fashion General Relativity as a geometric theory of gravitational effects. The curvature of space is specified within a framework of four dimensions. According to Bertrand Russell, the continuity of space-time expressed in Einstein's mathematics can be supplanted by a theory of discrete space-time that answers to all the same empirical data. Our basis should be well suited to modeling the "curvature of space" as a large-scale structural possibility in four dimensions.

Thus we are finished with our considerations of gravity, and we have yet to deal with electromagnetism and nuclear forces.

When we order our diagrams according to the number of individuals involved, we see a progression of increasingly complex forms. Each distinct diagram corresponds to a unique determinate "structure" as defined by isomorphism. Our basis provides a recipe for each unique structure as a list of ordered pairs of individuals. We have then, in principle, a systematic method of denoting any finite structure. It is such structure that we ascribe to the physical world in regard to its bare temporal succession, and for our strictly reductive purpose, we need no other language but that of "structure."

On our theory, an elementary particle is a sequence of individuals. That's all it can be. Therefore we look to the most primitive diagrams to find the most elementary "particles." We will confine attention to closed structures. Figure 10 shows all the valid closed diagrams that employ four individuals or less.

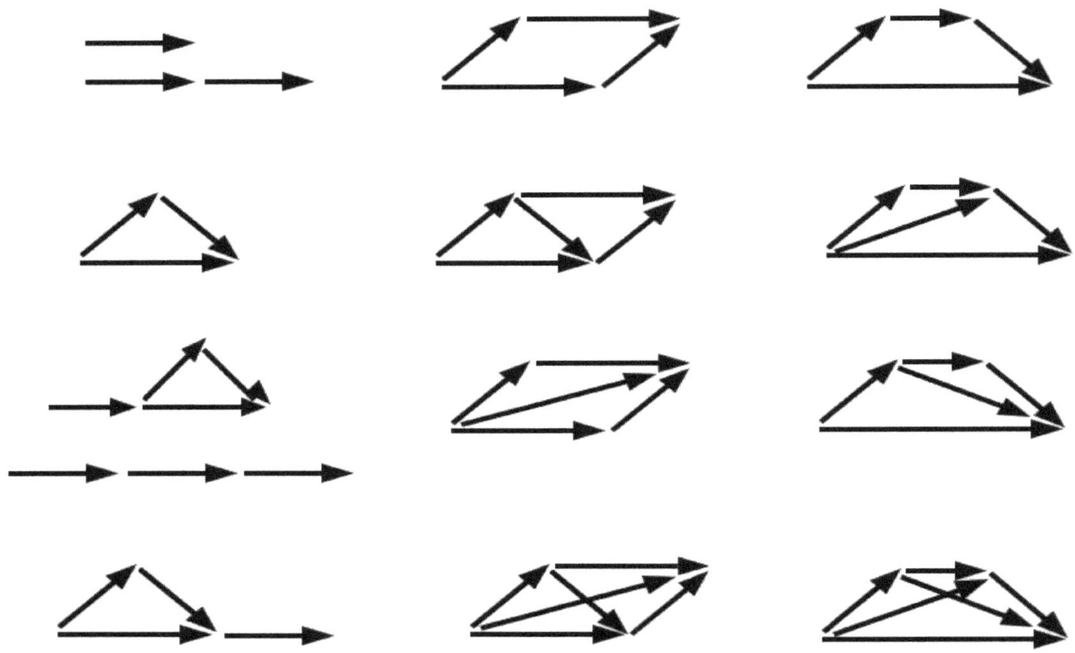

Figure 10

The diagrams are drawn in such a way that the earliest individual is on the left and the latest on the right. We can "bead" any of these diagrams together in a sequence, either with variety or by replicating one diagram, by joining the last junction of one to the first junction of the next. Any such serial stringing of beads will itself form another closed diagram. A "string of beads" can itself fork off and reconverge, yielding open diagrams when there is not a unique first bead and a unique last bead.

Let us suppose, for the purpose of discussion, that the photon has the simple structure of a line of arrows. We expect a photon to repeat its arrow indefinitely until it joins up with something else. The same goes for the other particles. The "law of inertia" in our theory seems to come down to something like this: "Any closed structure tends to repeat itself in succession until something else is encountered." Our basis provides this possibility but offers no reason or likelihood for it. We can't derive this law of physics from the basis. Whatever reason there may be for it must come from outside of physics. One may still hope to reduce the number of laws to a minimum with this theory.

A scientific theory is a prediction tool. Under this theory, prediction is purely a matter of projecting future structure on the basis of past structure. While our formal basis does not yield "the laws," any laws formulated in its terms will be assured of mutual consistency because the formal basis has the same consistency as the arithmetic of whole numbers.

For any closed diagram, like those above, we can take its bounded time interval as a provisional unit and calculate the relative durations and energies of its constituent quanta. Notice that among the "quadrilaterals" there are variations in energy for structures that have the same number of individuals. The "hollow" instances have the least energy, the ones with a single diagonal have more, and the last two are "full." Is a "full" particle more likely to "eject a photon" than its "hollow" counterpart? Is the hollow particle more receptive to the absorption of a photon? The ratio of individuals to arrows in a diagram may be an important parameter to keep track of.

Figure 11 shows a few of the many valid closed diagrams that have exactly five individuals. There are more than a hundred possibilities by my count, mostly due to the valid variations in "diagonals." The shapelier of the two pentagons gives rise to 48 variations. The number of arrows ranges from 5 to 10. Flipping and rotating a diagram cannot yield a distinct diagram. Flipping an arrow, however, sometimes does.

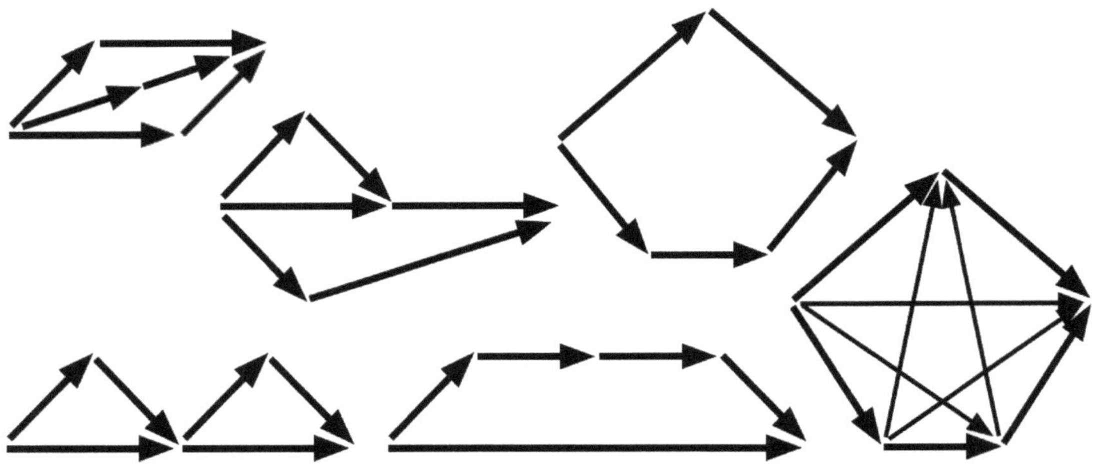

Figure 11

The “full pentagon” in Figure 11 is almost capable of forming a 4-dimensional latticework. If this structure is replicated to form a linear string, every junction of the “bead string” will look like Figure 8. “Clumping” together of closed diagrams is a secure method of trying to build up a “layered” model of the world from its highest frequency components to its lowest. The microstructure of succession provides the guarantee of consistency, and any determinate structure, no matter how large in scale, has a unique breakdown into individual arrows.

A hexagon has more symmetry than a pentagon, which restrains the growth of possibilities as we go from five individuals to six. There will certainly be a big increase, however. A photon can be appended or prepended to any diagram of “order five” to give a valid diagram of “order six,” which alone accounts for a doubling of possibilities. I believe I’ve drawn many “elementary particles,” but I don’t know *which* fundamental particles I’ve drawn. I’m not a physicist. I suggest the use of a computer for studying the burgeoning variety of these structures.

This is where I leave off. I am persuaded of the validity of this theory because a host of physical concepts heretofore considered fundamental appear to be reducible to a single ordering relation of utter simplicity. I commend this theory to the scientific community for further development— Carey Carlson, November 20, 2004

Addendum, November 29, 2004: The Electron

I meant to leave off, but I woke up at a strange hour and drew a doily pattern based on a hexagon with six diagonals. Excited to see that all interior nodes form a 4-dimensional lattice, with four ways coming and four ways going at each node, I figured this 2-D drawing of a finite contiguous history of 4-D time sequence would make a worthy cover diagram. I then wrote something for the back cover to describe the diagram. After admiring the diagram for several days, I realized I had been too timid. The diagram belies its own scale. We can identify the electron in it!

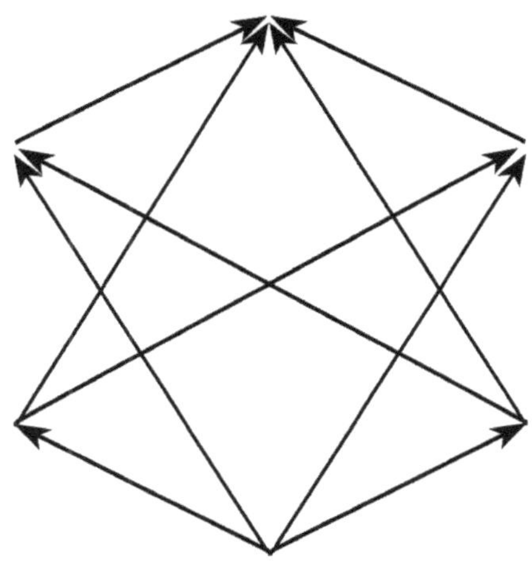

Figure 12: The Electron

The diagram of the electron is a closed diagram. If it repeats itself in a chain straight up the page, we have a single electron persisting in time. If it zigzags up the page, sharing two of its top three nodes in alternation, a "wider" cross section is presented to the overall direction of time, which is roughly "upward" in the diagram. This represents a helium atom. The next wider ribbon up the page would be a 1-2-1-2 pattern of "electron tower." This much widening has been interpreted as "going to a second shell" in three dimensions, and we have lithium. Then there is a 2-2-2-2 zigzag: beryllium. Then comes the 2-3-2 tower of electrons depicted by the cover diagram: boron. And so on through the "electron cloud" structures of the periodic table of elements.

So now we have, in Figure 12, the quantum structure of the electron to look at. Six individuals in ten succession pairings. 10 quanta of energy. Since the electron has succumbed to analysis, it is "elementary" no longer. Sticking with the rule of thumb that "simplest is most likely," the simple electron is most likely the basic building block of our four dimensions. (I'm ignoring the nucleus, and the higher frequency patterns of its particles; these latter must make termination with the nodes of the electron in a systematic way.) Recall the discussion that sprang from consideration of a "hole" in the otherwise uniform 2-dimensional lattice of Figure 6. We start with Figure 6 and "cut away" the bulk of the diagram to obtain Figure 7, a diagram of two particles engaged in spatial separation. Similarly, we may now start with a uniform lattice of four dimensions, built entirely from electrons, and carve away everything that doesn't look like the four-dimensional world we inhabit. We obtain as much "empty space" as we see fit, leaving a skein of electrons to represent the solids, liquids, and gases that we find around us in haphazard distribution.

The electron I've diagrammed (more accurately, one *cycle* of an electron) is not "full." It can "take additional arrows." It can take one straight up from earliest to latest, for example. It can take a vertical along each side. And it can take a horizontal, or two. Then it's full. When electrons form a phalanx in their march through time, they share vertices, economizing on the number of individuals and the number of arrows per electron. The lower energy level contributes to the stability of a given phalanx, or "electron shell."

I read something today about the "magic number" 137 involved in the structure of the atom. It is supposedly the ratio of an electron's velocity in the inner orbit to the speed of light. Although electrons don't "speed around a nucleus" in this theory, something has heretofore been interpreted that way. If 137 is indeed an important integer, I expect its significance will be revealed in a fairly simple arrow diagram.

Addendum, December 4, 2004: The Big Bang

Upon further reflection, it appears that the "Big Bang" has been misconstrued as a singularity marking the beginning of time, when it is more likely that our system of temporal succession has no beginning. As we consider earlier and earlier stages of the universe, we conceive a spatially shrinking universe, converging quite naturally according to our spatial intuitions to a point-like minimum of spatial extent. But the clarified situation involves higher frequencies, and the "shrinking of space" is a pure consequence of these increasingly higher frequencies. A recent issue of Discover magazine featured a marble-sized sphere on the cover graphic that represents the spatial size of the universe at t = 10e-34 seconds. Yes, space shrinks toward zero as all frequencies increase, but clarity on the subject again demands that we relinquish unanalyzed spatial intuitions. The integer exponent in the equation is heading toward larger integers (higher frequencies) as we delve into earlier stages of the universe. Nothing in the basis of our theory suggests that frequencies should have an upper limit, unless we restrict consideration to a bounded region. Nothing suggests that the exponent in the equation will, with further knowledge on our part, attain some satisfactorily high integer and reach a "glass ceiling." Physicists of the distant future, should they be likewise captive to their spatial intuitions, might well conclude that our age was one of unbearable heat, a marble-sized universe suffering from excruciating proximity to a cataclysmic Big Bang.

Addendum, December 23, 2004: The FSC

The left side of Figure 13 is the cover diagram, except that it is outfitted with a first and last electron to make it a closed diagram. In addition, displaced rightward for ease of discernment, are all possible pairings of vertically aligned nodes of the closed diagram. There are 137 arrows altogether, 78 on the left and 59 on the right. The integer 137 is sometimes referred to as the "fine structure constant." On this theory, we are indeed looking at the fine quantized structure of a closed time sequence, which I now take to be one cycle, or time period, of a helium atom.

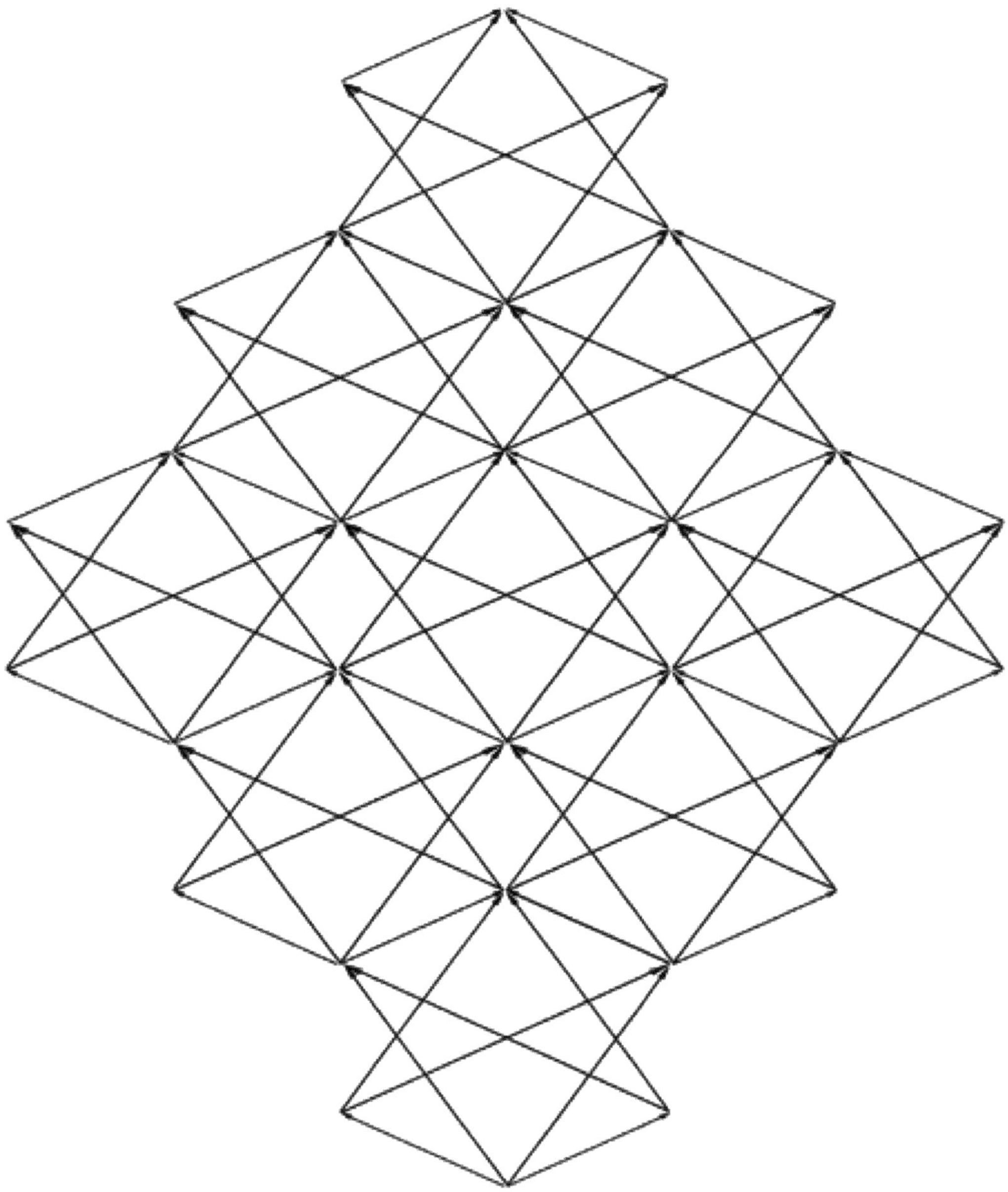

Figure 13

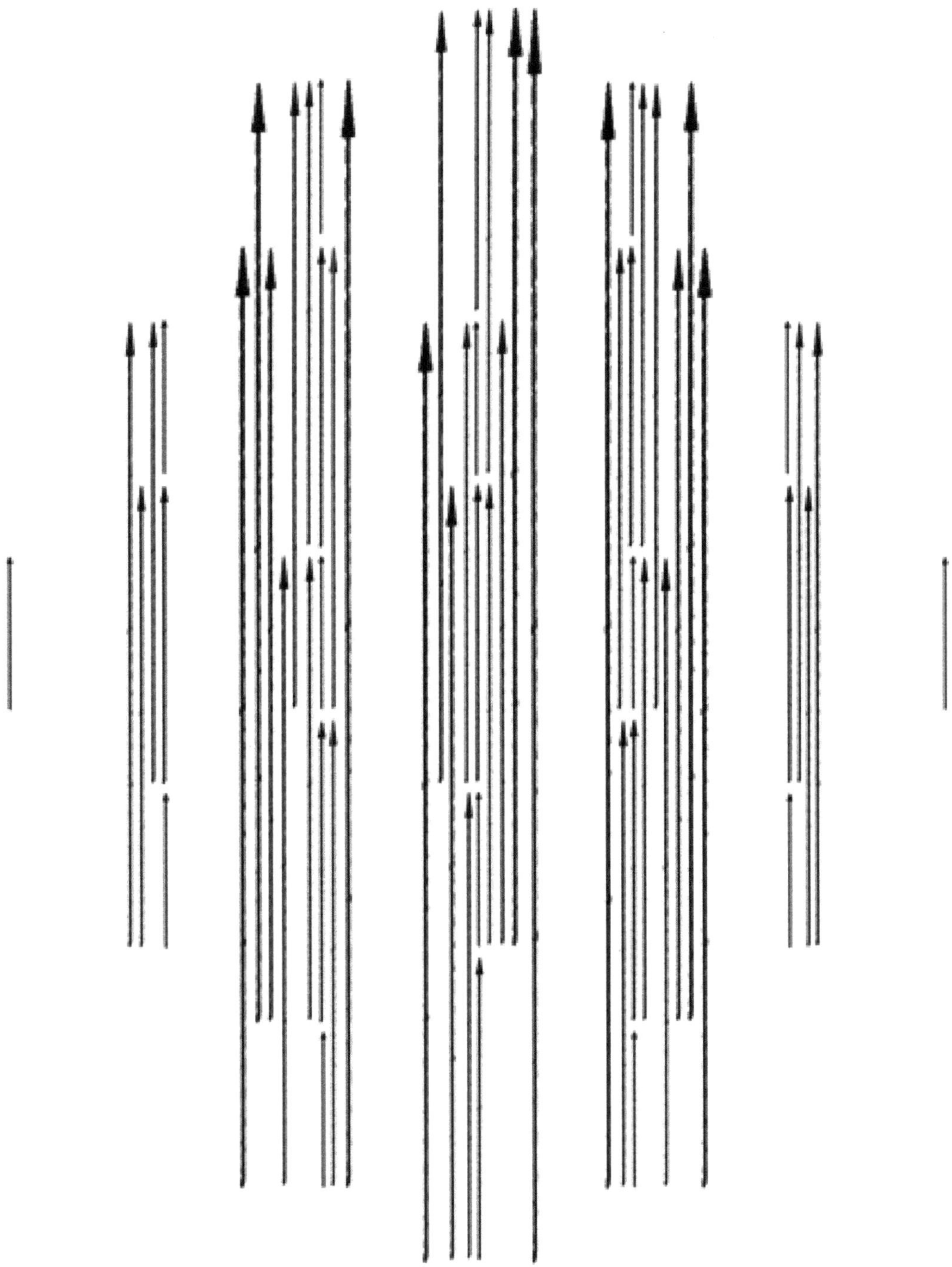

If indeed this is the ultimate analysis of the fine structure constant, I draw two immediate conclusions. First, the closed diagram is more important, in modeling nature, than I first realized. My "periodic table" assignments were too hasty. A stable atom seems to demand a periodic finish to its cycle, and more symmetry too, than the "wobbly towers" I was imagining. Secondly, all verticals that can occur do occur, an apparent law of nature, and quite unexpected to me, I assure you.

By maintaining symmetry in the 2-D geometry of my hexagons, the "axes" and "parallels" of the above diagram *do have meaning* for the structure of the purely topological time sequence. The 59 "verticals" make up the dominant of six discernible "time axes" of the helium atom. The 78 quanta on the left align into 4 axes, and I've drawn no "horizontal" quanta, the sixth axis. Then there are other "off-axis" possibilities, none of which have been drawn in. The verticals make a sort of histogram of frequency components of the helium atom's "on-axis" quanta, organized into a center, a spacing, a shell (not really a shell for "two electrons"), and some "sidebars." In any case, it seems quite remarkable that on-axis quanta "fill in" all possibilities to make 137, as though nature has an inexhaustible store of quanta to satisfy such possibilities.

A Topological Derivation of Atomic Spectral Data

My guiding assumption is that physics should be reformulated, from the ground up, in terms of the topology of discrete time sequences. My first effort in that direction (A Theory of Everything for Physics, 2004, Prism Publishing) culminated in a proposed quantum structure for the electron and its cloud formations. In the diagram below, a chained sequence of 36 electron cycles runs up the left side, representing a free electron persisting through time. To the right of the free electron is a stable hydrogen cloud, persisting through nine of its cycles. To the right of that is a stable helium cloud through four of its cycles. Each of those sequences is iterated enough to incorporate 36 component electron cycles. On the far right, a hydrogen cloud absorbs a photon, which boosts the sequence to a helium cloud formation. Soon after that, a photon is emitted, and the sequence reverts to a hydrogen cloud. The frequency of the photon (the first spectral line of the Balmer series for hydrogen) is equal to the frequency of the hydrogen cloud (f_2) minus the frequency of the helium cloud (f_3). The other Balmer lines, and the spectral line series of other atoms, are obtained similarly, with the progression of cloud structures supplanting the "orbits" in Bohr's model of the atom.

The spectral line wavelengths are generated by Bohr's formula, $R(1/k^2 - 1/n^2)$. "R" is the Rydberg constant. We adopt the electron cycle, a topological construct, as the unit of energy. We postulate that n electron cycles establish the same time interval regardless of the cloud structure, as suggested by the scaling employed in the diagram. The wavelength of a free electron is 1 and its frequency is 1; the wavelength of hydrogen is 4 and its frequency 1/4; the wavelength of helium is 9 and its frequency 1/9, and so on. Thus, the frequency of the photon depicted in the diagram (the inverse of its wavelength) is 1/4 – 1/9, satisfying Bohr's formula and my own equivalent formula, $f_2 - f_3$. (If the Rydberg constant does not resolve to 1 with the change to non-arbitrary units of measure, our results are correct but for a scale factor.)

Note: The scaling suggests a "slowing of time" with increasing cloud size, a basis for gravity.

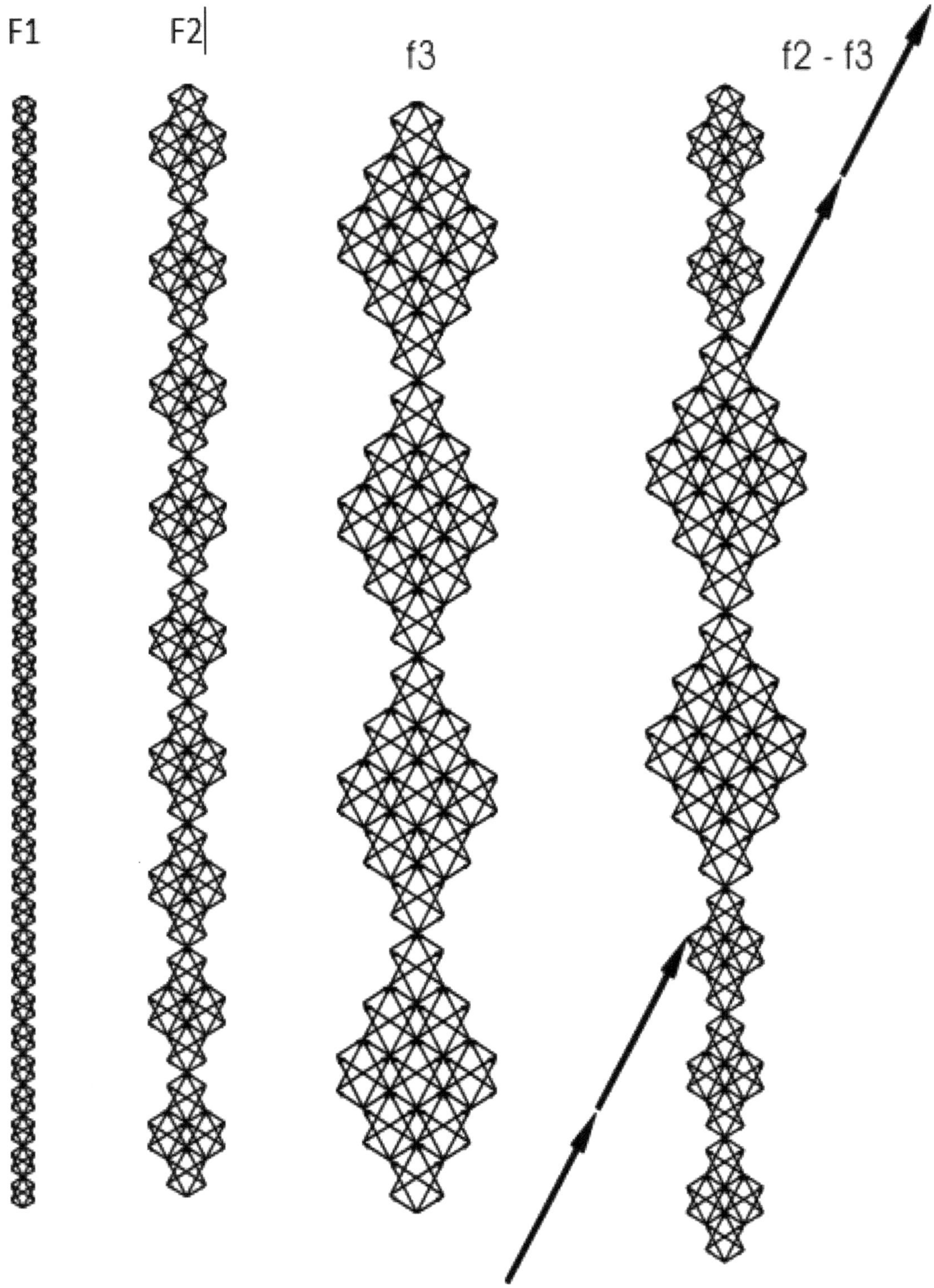
F1
F2
f3
f2 - f3

CAUSAL SET THEORY AND THE ORIGIN OF MASS-RATIO

Carey R. Carlson, June 27, 2010

Abstract

Quantum theory is reconstructed using standalone causal sets. The frequency ratios inherent in causal sets are used to define energy-ratios, implicating the causal link as the quantum of action. Space-time and its particle-like sequences are then constructed from causal links. A 4-D time-lattice structure is defined and then used to model neutrinos and electron clouds, which together constitute a 4-D manifold. A 6-D time-lattice is used to model the nucleons. The integration of the nucleus with its electron cloud affords a calculation of the mass-ratio of the proton (or the neutron) with respect to the electron. Arrow diagrams, along with several ball-and-stick models, are used to streamline the presentation.

Ideas must have formal expression to become part of physics. In order to formalize Whitehead's idea of "temporal succession," a temporal successor relation is required. Ordered-pair notation suits the purpose. A single step of succession is denoted by a single ordered pair: (a,b). The arguments denote individual moments of time. A two-step sequence is denoted by a set of two ordered pairs with one member in common: {(a,b), (b,c)}. Any finite temporal structure can be denoted by such an interlinked set of ordered pairs.

Next, we use the successor relation to define "earlier than":

1. For any two moments, if (a,b), then a is also earlier than b.
2. For any three moments, if a is earlier than b, and b is earlier than c, then a is earlier than c.

We then characterize time order with a postulate:
No moment can be earlier than itself.

We now set out to reconstruct physics from time order alone. We purposely neglect providing any axiom of infinity. Any construction we perform is thus confined to a finite number of ordered pairs. Any subsequent analysis into whole-and-part will therefore arrive at logical primitives in a finite number of steps. The only primitives are the moments, along with instances of the successor relation. As a consequence of this whole-and-part finitude, there is no infinite divisibility of a time interval, no continuous manifold, and no calculus in the theory.

A graphic arrow, with its directional asymmetry, serves just as effectively as an ordered-pair expression to depict a temporal transition from one moment to another. We can use the argument letters to label the endpoints of the arrow:

We can depict a time series of three moments as follows:

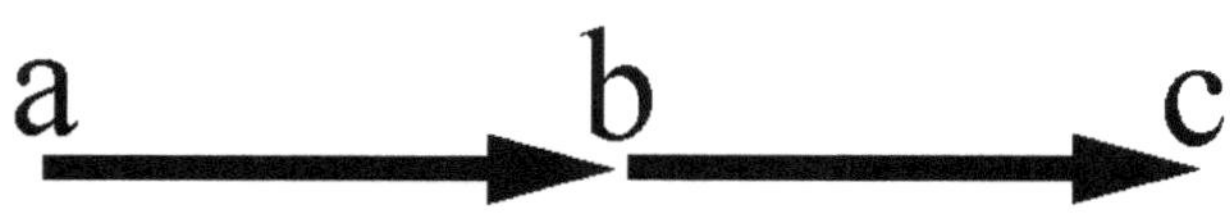

We can construct a time series of any finite length that we like, but we can also construct time-sequence possibilities that are not serial. There are four distinct ways that three moments can be arranged in chronological order. These are diagrammed below, along with one "impossible figure" that violates chronology protection.

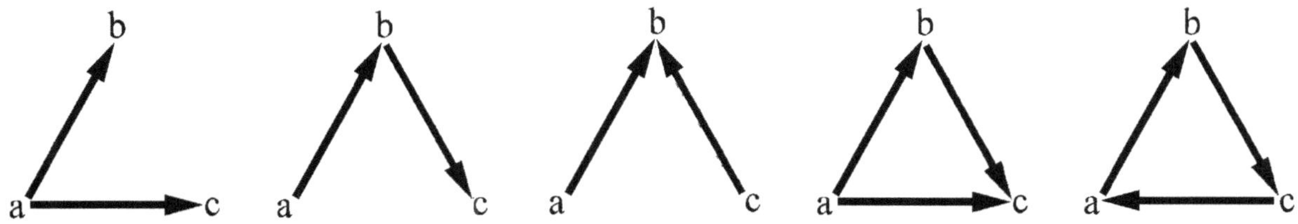

The offending diagram is the one on the far right, because each moment is earlier than itself. The first four diagrams are fine, and we assign them common names, from left to right: fork, series, convergence, triangle. Each diagram gives rise to variations if we swap the labels around while leaving the arrows undisturbed. It is structure, as defined by isomorphism, that remains undisturbed by the label swapping. The variety of structure provides the variety of physical entities in this theory. The formal basis yields a limited variety of structural possibilities for a finite number of moments. That "limited variety of natural kinds" provides a basis for the application of probability theory. However, the theory is founded on the enumeration of all structural possibilities, starting with the simplest ones shown above, and probability plays no part in this.

There is only one distinct diagram with exactly one arrow. There are three distinct diagrams of two arrows, counting the fork, series, and convergence. The series surprises no one, since it is nearly an automatic assumption that time order is strictly serial. The fork and convergence contravene that assumption. Time order and causal order are conflated in this theory, such that "temporal succession" and "causal succession" are interchangeable terms. Any fork in the time diagrams thus depicts a single cause with multiple effects, and any convergence depicts a single effect with multiple causes.

Causal Set Theory

Our formal basis is the same as that of causal set theory. The choice of terminology to accompany the arrow diagrams is flexible. I favor the terms "moment" and "transition" for the two primitives of the theory. Causal set theorists use the terms "elements" and "causal links," respectively, for the same two primitives. In the diagram notation, these terms correspond to the junctions (or nodes) at which the arrows meet, and the arrows themselves.

The Origin of Mass-Energy

The triangle diagram consists of two locally separable paths that begin together and end together. A relative frequency ratio is formed, which compares a 2-step path with a single-step path that traverses the same time interval. Without the structural feature of forking paths that rejoin, relative frequency does not arise. The time triangle is thus the simplest causal set to feature an inherent relative frequency. We can construct a diagram of relative frequency for any rational number. The following diagram features a 2:3 relative frequency ratio:

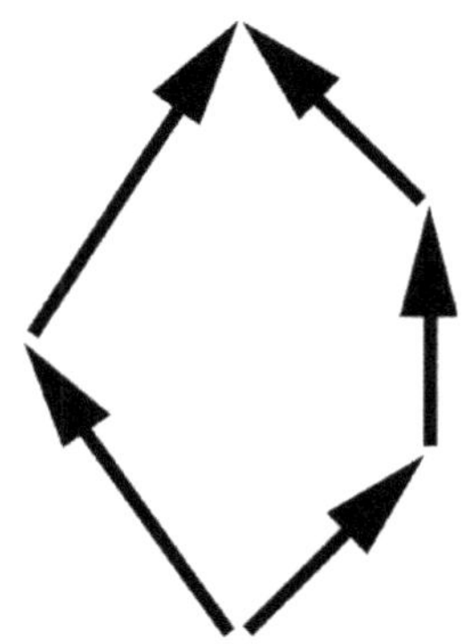

Relative frequencies serve this theory as relative energies in accord with Planck's E = hf. Two energy values for comparison, E1 to E2, correspond to hf1 and hf2 respectively. Thus, E1/E2 = f1/f2. (Planck's constant drops out.) This suggests that an energy ratio is just a frequency ratio, and that energy only arises, and only has meaning, in ratio-comparison form.

In regard to frequency, all that we can glean from the above diagram is the ratio of arrow-count for two locally separable time pathways. The 2:3 ratio measures the energy ratio of the one path to the other. The individual arrows are the countable units that yield the integer components of the numerical ratio. By that consideration, the individual arrows of our diagrams depict the individual quanta of quantum theory. Energy is packetized in quantum theory. Here, the flow of time is itself packeted into discrete transitions, and these serve the theory as quanta of action.

Empirical measurement procedures are also confined to ratio-comparisons, as many astute theorists have pointed out. We measure one mass against another, one spatial span against some chosen yardstick, one time interval against some de facto unit of time, or one frequency against another, as with the interferometer. The structural definition of relative frequency in this theory provides the ontological ground for the ratio-comparison constraint that characterizes practical measurement techniques.

We tend to employ an arbitrary unit of measure, then forget that we are doing so, and think of energy as having an absolute value as a scalar quantity. We then wonder about "the origin of mass." The answer is not to be found in the origin of a scalar quantity, but in the structural origin of discrete ratios that arise in patterns of sheer temporal succession. This account of the structural origin of mass-ratio will bear fruit in the developments ahead. We shall first construct the electron, then the proton, and then derive the ratio 1:1836 as their mass ratio. The derivation of the ratio proceeds without obtaining any scalar mass value for either particle alone, illustrating the point that has just been made.

The reciprocal of frequency is wavelength. Higher frequencies equate to shorter wavelengths. In temporal terms, wavelength is a measure of duration, or time period. Higher-frequency paths consist of shorter-period quanta. Frequency ratios and their reciprocals thus measure energy and wavelength ratios, respectively. We obtain the numbers that physics requires for frequency and wavelength without invoking either waves or particles.

Defining "closed diagram" and "bounded region of time"

Some diagrams have a single earliest moment and a single latest moment. Such a diagram shall be termed "closed." A diagram that is not closed is termed "open."

A "bounded region" is defined as the ordered complex of all the quanta located between two defining moments, one earlier than the other. The two defining moments are the earliest and latest moments depicted in the closed diagram of the region. Such causal boundedness characterizes what can be learned by the scientific method.

The causal patterns that play out in the runtime of an experiment are localized by two bounding moments: the moment of initiation, when the causal laws being tested are triggered into action, and the moment of completion, when the outcome of the experiment is known. Between those two moments, nature enacts a causal sequence that culminates either at the predicted outcome or at some other outcome.

Closed diagrams will also assume special importance in the modeling of stable particles and the calculation of their mass ratios.

Our causally connected universe, we must suppose, corresponds to one elaborate arrow diagram. What are its highest and lowest frequencies? In a bounded region, the frequency range is capped at both ends. This ensures that energy density is everywhere finite. In the region of a high-energy experiment, we produce higher frequencies than those of ordinary matter in a livable environment. Confining our scope to the latter, the up/down quarks contain the quanta of highest frequency and shortest time period. That marks the high end of nuclear frequencies in ordinary matter. From there, a ladder of discrete lower frequencies extends to the frequency of a free electron. The EM spectrum begins there and extends to lower and lower frequencies, finally fading out of detection due to increasingly feeble quanta. The low end of the frequency spectrum, corresponding to quanta of the longest time period, is limited, if at all, by the age of the universe. (The age of the universe is not necessarily finite in this theory.)

The diagram notation is interchangeable with the ordered-pair notation we began with. Any diagram can be labeled at its nodes, and each arrow is then translated into an ordered pairing of the labels at its endpoints. The theory can be expressed entirely as the combinatorics of ordered-pair expressions. That is the safeguard against over-interpretation of the diagrams. The shape and size of the diagrams are irrelevant artifacts of planar geometry that indicate nothing in the domain of reference. A diagram specifies nothing but time order. That said, we can streamline the presentation by relying on the diagrams for the more intuitive recognition of structure that they provide.

Confining attention to closed diagrams, here is a complete set of the closed diagrams that employ four moments or less:

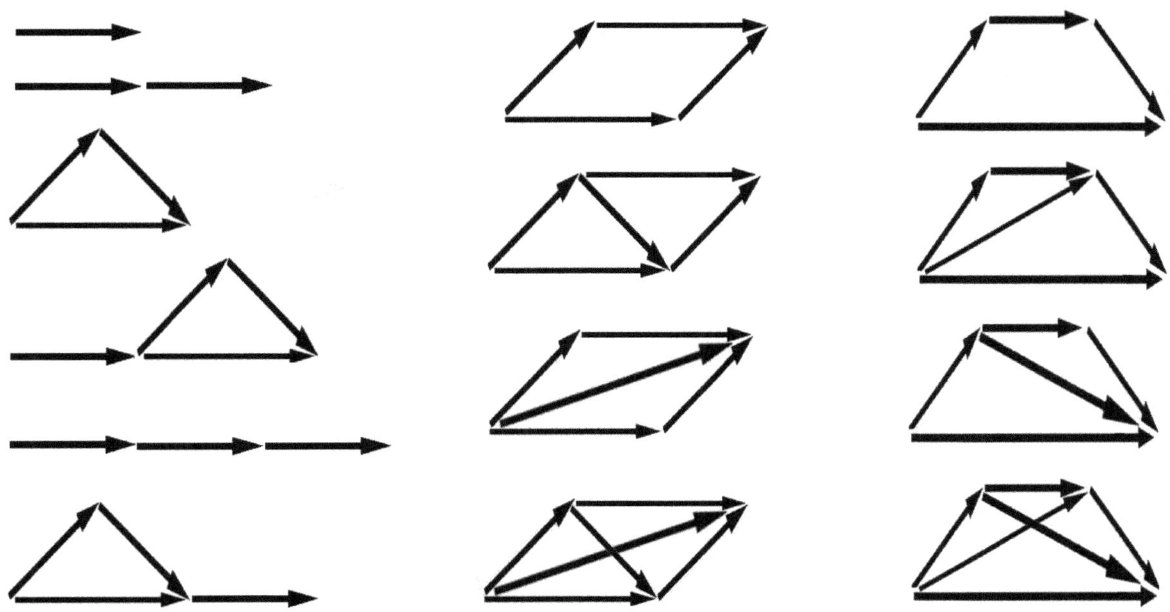

The top-center diagram I call "the primitive diamond." It can accommodate one or two additional arrows without any increase in the number of moments, as depicted in the center column. Looking down the center column of diagrams, each finite sequence is a distinct chronological structure as defined by isomorphism; the absence of a pairing connection between two particular moments is no less important in denoting structure than the presence of a pairing connection.

Any closed diagram, including the ones shown above, can serve as a basic repeating "cycle" for constructing further closed diagrams, by defining the latest moment of one cycle to be the earliest moment of another such cycle. This is called "chained repetition." For instance, the primitive diamond can propagate indefinitely in chained repetition. Such a propagating sequence of characteristic cycle-structure will replace the notion of "a particle persisting through time." The particle-like sequence will have its characteristic De Broglie wavelength and frequency. It turns out that the most prevalent stable particles in physics are modeled by chained repetition of the simplest, most symmetrical time patterns. The primitive diamond and its variations must model something very basic, such as polarized light, but I haven't studied that possibility.

We shall skip past the diagrams that have exactly five moments, in order to focus on the following arrangement of six moments, connected by ten arrows, which I call "the hex cycle."

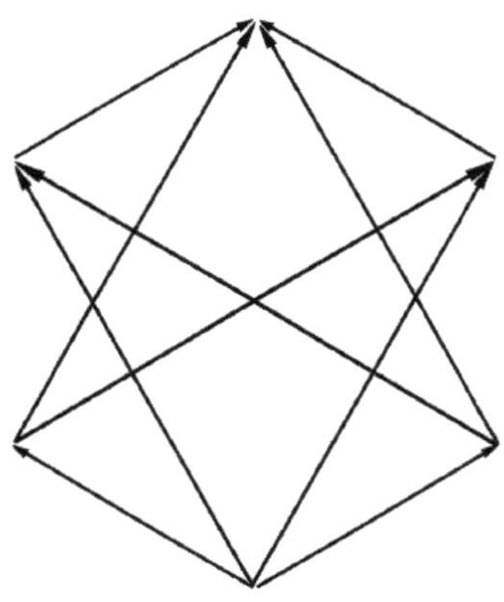

The hex cycle diagram is closed and highly symmetrical. In chained repetition, it will model a neutrino. Equally important, the hex cycle serves as the basic "cell" from which to compound the cell-like structure of four-dimensional time. I've drawn the hex cycle on the points of a regular hexagon so that it can be tiled graphically with others like itself to make an extendable four-dimensional time lattice.

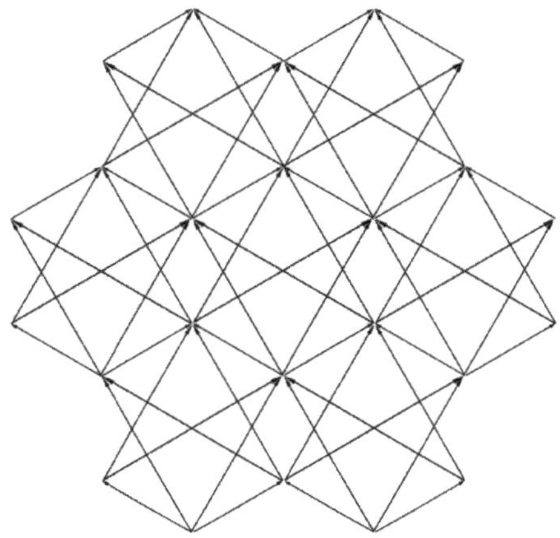

In the above diagram, six hex cycles are ringed around a center one. Additional outer rings can be added to extend the tiling pattern uniformly in all directions without limit. In the absence of a continuum, dimensionality is a matter of counting the arrows that meet at each node of a regular pattern. Notice that every interior node is at the intersection of four time axes, with four arrows arriving and four departing. This structural plan for an extendable four-dimensional manifold will serve to replace "space-time."

The "Honeycomb Series" of Closed Diagrams

We can also define a series of diagrams of increasing extent by compounding hex cycles in the following manner, which produces only closed diagrams.

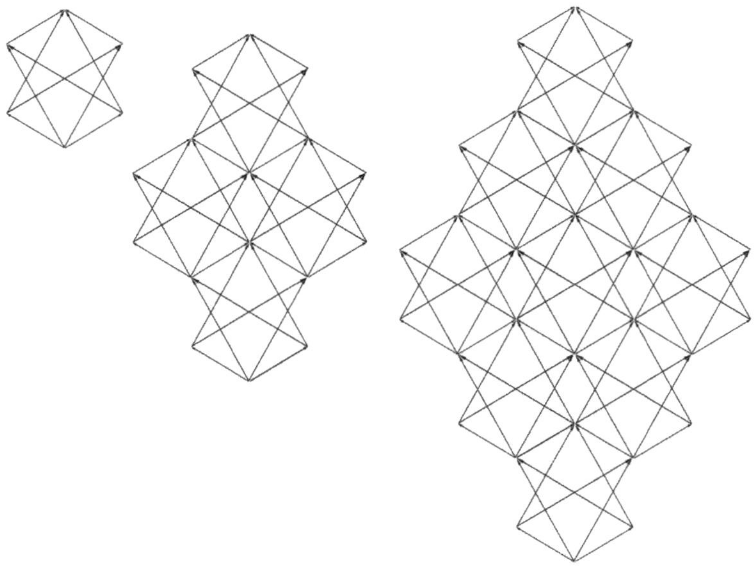

In the above series, as in the previous tiled diagram, every interior node is at the intersection of four time axes. Every quantum belongs to one of the four time axes. We shall call these quanta “lattice quanta,” because they form a localized portion of a four-dimensional time lattice. Choosing a sufficiently large “n” as the nth member of the above series, we can obtain a bounded region of four dimensionality suited to span a cosmic scale. Such a four-dimensional manifold, like anything in this theory, is made of quanta. The four-dimensional volume of any subregion is measured by the number of component hex cycles.

We are reformulating Special Relativity to employ four time axes all alike, rather than three of space and one of time. All coordinates are then real-valued, and we dispense with imaginary numbers. Also, there is no need to invoke an axiom for the limiting velocity of light. That limit is a logical consequence of excluding instantaneous spatial relations from the theory. A spatial relation is, in effect, a simultaneity relation. The exclusion of primitive spatial relations from physics is the simplest way to implement Special Relativity and its “breakdown of simultaneity.”

Quanta of momentum and charge

It is charge quanta that distinguish electrons from neutrinos. See the following figure, which shows the additional quantum locations afforded by the hex cycle.

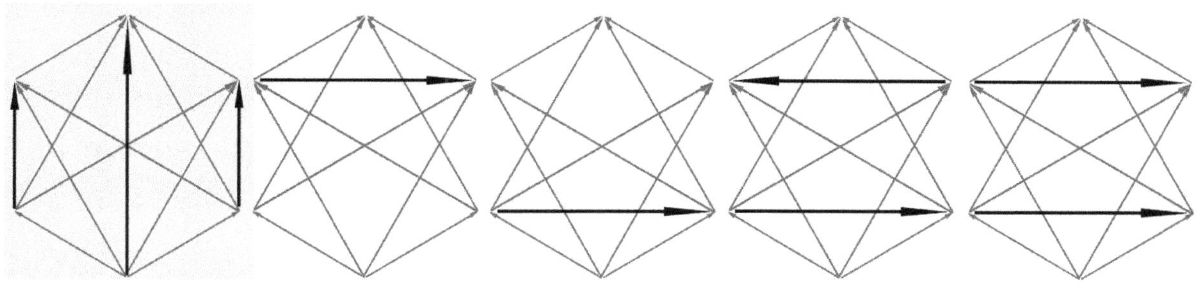

Placing the greatest importance on the hex cycle as the basis of our four-dimensionality, I have preserved its integrity as the template of all five diagrams. Variations are obtained by selective inclusion of the additional quantum possibilities afforded by the six moments of the hex cycle. In the leftmost diagram, I have drawn quanta of forward momentum. That hex cycle has a major-axis quantum, the center arrow, which transitions directly from the earliest to the latest moment of the cycle. Such a quantum has the greatest duration and the least energy of any quantum in its cycle. It occupies the axis of bilateral symmetry inherent in the hex cycle. I have also drawn both of the other "vertical" quantum possibilities, thereby preserving the bilateral symmetry. The "parallelism" of those three quanta is a topological feature of the hex cycle, and not just an artifact of the geometry of regular hexagons that I've employed in my drawings. The major axis and its parallels define the axis of proper time intrinsic to the hex cycle.

The other four diagrams include either or both of the charge quantum possibilities. The presence of any charge quantum breaks the bilateral symmetry about the major axis of the cycle. Flipping or rotating an entire diagram does not affect its structure, so the four diagrams above exhaust the charge possibilities of a single hex cycle. If we were to include all three verticals plus two horizontals, two charge quanta, in one hex cycle diagram, the diagram would be full. More moments would be needed to accommodate more quanta within the region of such a cycle. The limited capacity of the hex cycle for additional quanta serves the same function as the Pauli exclusion principle.

Symmetries of the hex cycle correspond to the symmetry rules governing charge-conjugation, parity, and time reversal. For example, a hex cycle of ten arrows has mirror symmetry about its major axis (parity). Also, if we reverse the direction of every arrow in a hex cycle diagram, we end up with the same diagram we started with (time reversal). There are three further arrangements for quanta in the proper-time axis of the hex cycle that serve to explain the spin parameter of QM:

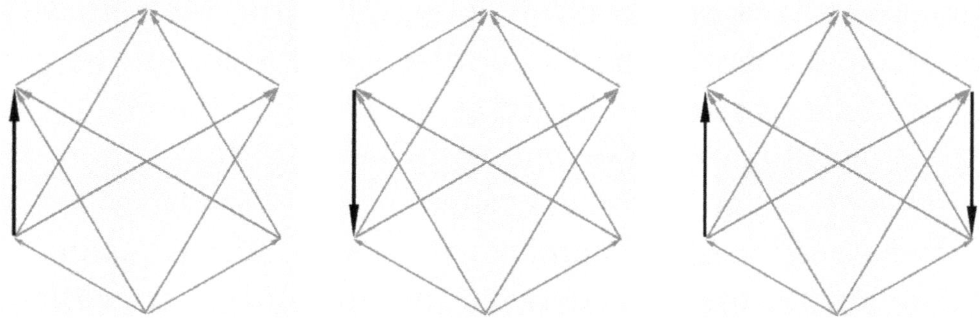

The vertically aligned arrows are in the proper-time axis, whether they point up or down in these hex cycle diagrams. Chronology protection precludes two down-arrows in a single hex cycle, because a four-step closed loop would be formed. Only one down-arrow can be accommodated. A down-arrow constrains the orientation of charge quanta that can accompany it. For instance, the hex cycle on the right can accommodate up to two charge quanta, but given the orientation of the up/down arrows in that cycle, the left-right direction of each charge quantum is dictated by chronology protection. Thus, there is a type of "handedness" to a hex cycle that includes down-arrows.

Bohr's Formula

The chained repetition of any one diagram in the "honeycomb series" depicts a mode of neutrino propagation. Electrons and their cloud formations have the same structure as the neutrino formations, except that electron clouds are occupied by charge quanta, while neutrino formations consist entirely of lattice quanta. The chained repetition of a single arrow models the photon of radiant energy. The justification for these claims is given in the next illustration, which models Bohr's formula for the "spectral fingerprints" of the atoms.

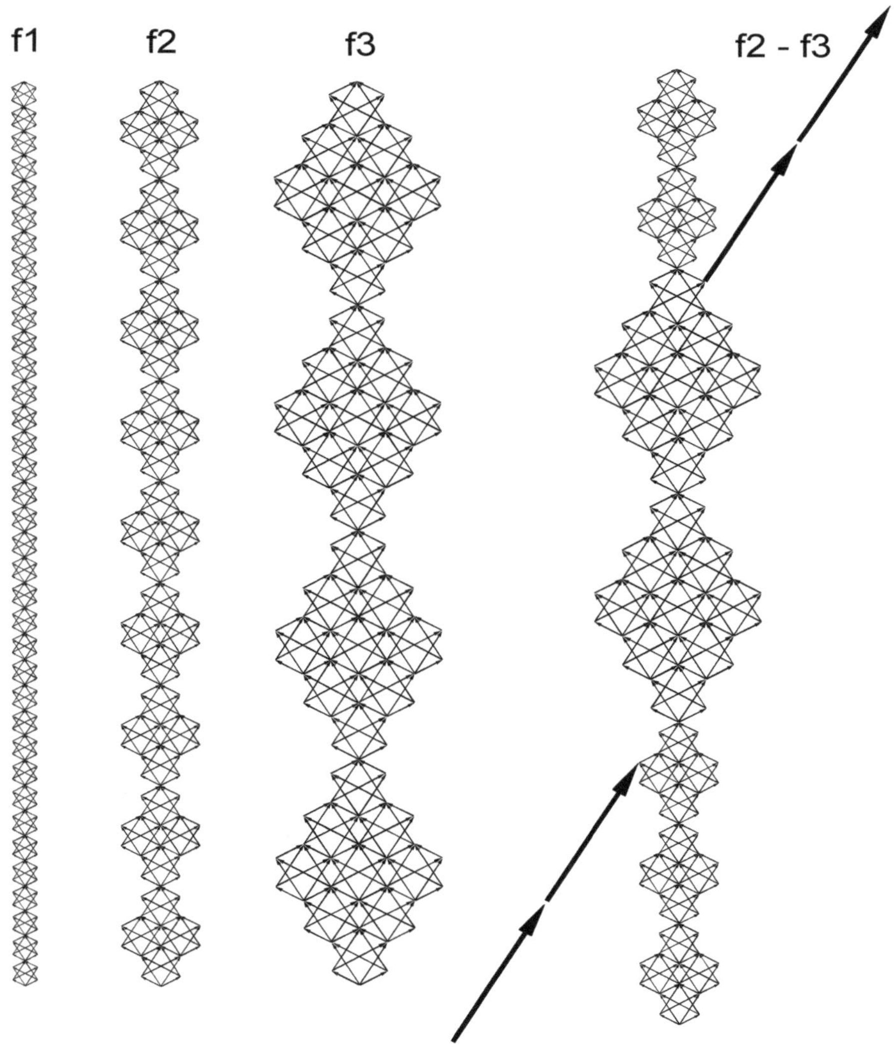

I have drawn the hex cycles as neutrino cycles, without any charge quanta. As drawn, they depict modes of neutrino propagation. If we populate the hex cycles with charge quanta, we get electron clouds. In that case, we would have, from left to right, a free electron, a hydrogen cloud, a helium cloud, and finally a hydrogen cloud sequence disturbed by an encounter with a photon. The first three diagrams mark the start of a progression that continues in step with the periodic table. The stable clouds feature uniform cycles of uniform frequency. The cloud disturbed by the photon exhibits a modulation of cycle-pattern and frequency. The frequency of the photon, either incident or emitted, is the difference in frequency of the two cycles involved in the modulation. This scenario, generalized to the whole series of cloud possibilities, yields Bohr's formula for the spectral wavelengths of photons absorbed and emitted by the atoms.

Counting the hex cycles of a sequence gives a rough measure of the total energy of that sequence. Each sequence is scaled so that thirty-six hex cycles span the height of the diagram. Thus, the same amount of energy plays out in the same amount of time for any of the cloud formations. That conjugate relation between time and energy holds quite generally, even when the nuclear energies are taken into account.

The spectral line wavelengths are generated by Bohr's formula, $R(1/k^2 - 1/n^2)$. "R" is the Rydberg constant. We adopt the electron cycle, a topological construct, as the unit of energy. The same electron cycle also provides a natural unit for duration and frequency. We postulate that n electron cycles establish the same time interval regardless of the cloud structure, as implemented by the progressive scaling employed in the series of diagrams. The wavelength of a free electron is 1 and its frequency is 1; the wavelength of hydrogen is 4 and its frequency is 1/4; the wavelength of helium is 9 and its frequency is 1/9, and so on. Thus, the frequency of the photon depicted in the diagram, the inverse of its wavelength, is 1/4 – 1/9, satisfying Bohr's formula and my own equivalent formula, f2 – f3.

Bohr's formula encapsulates a wealth of spectroscopy data. An explanatory model of the underlying structure of electrons and their behavior should not be more complex than the empirical data it is supposed to explain. The model represents theory, while the tidy formula represents the data that the theory is concocted to explain. In contrast to Bohr's model, we now have a model of electron-photon dynamics that is nearly as simple as the formula that encapsulates the data. It is a model consisting of quantum schematics that can be diagrammed precisely. With this model, the Rydberg constant resolves to unity, because the electron cycle provides the unit of mass and the unit of charge, while the constants "c" and "h" resolve to unity by more general considerations. We're left with the diagrams themselves, in which the electron appears as the "realtime clock" that sets the base frequency for the EM spectrum.

Time Dilation in the Foregoing Illustration

On this theory, the "slowdown of time" is the origin of gravity, rather than a side-effect of General Relativity. Progressive scaling has been applied to each successive cloud formation in the atomic series. Admittedly, this was done in order to model Bohr's formula. The energy inherent in a closed-diagram sequence is evidently tied to the metric of time steps in that sequence, so that the two are co-determined. The cumulative effect of hex-cycle expansion with larger cloud formations is the systematic deformation of the four-dimensional manifold according to local mass density. This constitutes a discrete version of General Relativity's "curvature of space-time."

The inverse-square laws, for both gravity and electromagnetism, are directly related to the inverse squares of Bohr's formula. The reason is this: the four-dimensional manifold that's been called "space-time" is nothing other than the aggregate of all neutrino/electron formations. We have seen that neutrinos and electrons are formed with the structural principle of an extendable four-dimensional time lattice. All such sequences connect to one another at some point in their past and at some point in their future, because the theory is built up entirely from causally closed regions. (That confines theoretical constructions to the scope of the scientific method, since the closed-region experiment is the only means of subjecting theory to empirical confirmation.) The connected "skein" of neutrino/electron formations in a closed region accounts for the four dimensionality of Special Relativity, without invoking any additional homogeneous background. The gaps that break up the uniformity of the manifold delineate the neutrino/electron formations that comprise what there is of that manifold.

Since Bohr's formula characterizes the metric of the "neutrino foam," it characterizes the underlying metric of space-time itself. That is why the inverse squares in Bohr's formula dictate the inverse-square laws that govern separation/approach dynamics.

One expects linear incrementing from an independent time axis. But proper time is not an independent axis; it is a resultant axis of four-dimensional propagation. The hex cycles establish a proper-time axis in conjunction with four axes of lattice quanta. We measure time along such a proper-time axis, which increments as an integral component of our local four-dimensional lattice propagation, such that cycles of equal energy transpire in equal time periods. If we fail to take into account the disparate proper-time metrics of locally isolated inertial systems, we attribute the mutual approach or separation of such systems to the deflecting work of forces that obey an inverse-square law.

The Fine Structure Constant

Returning to the "honeycomb series" of cloud cycle formations, let us use N to index the members of that series, so that N = 1 for the cycle of a free electron, N = 2 for the hydrogen cloud cycle, N = 3 for the helium cloud cycle, etc. The following formulas then apply to cloud-cycle N:

The number of hex cycles is N^2.

The number of lattice quanta is $8N^2 + 2N$.

Defining S(N) as the sum of squares from 1 to N, the number of locations for quanta in the proper-time axis is $4S(N) + N^2 - 2N$.

Testing out the latest formula, for N = 1, we get 4 × (1) + 1 − 2 = 3, which is correct for the single hex cycle. For N = 2, we get 4 × (1 + 4) + 4 − 4 = 20, which is correct for the hydrogen cloud cycle. For N = 3, we get 4 × (1 + 4 + 9) + (3 × 3) − (2 × 3) = 59, which is correct for the helium cloud cycle.

Using the above formulas, for any given cloud cycle N, we can add the number of lattice quanta to the number of proper-time locations. For N = 3, we get 78 + 59 = 137. The next drawing shows a helium cloud cycle with the two components separated out. The proper-time quanta are displaced to the right for ease of discrimination.

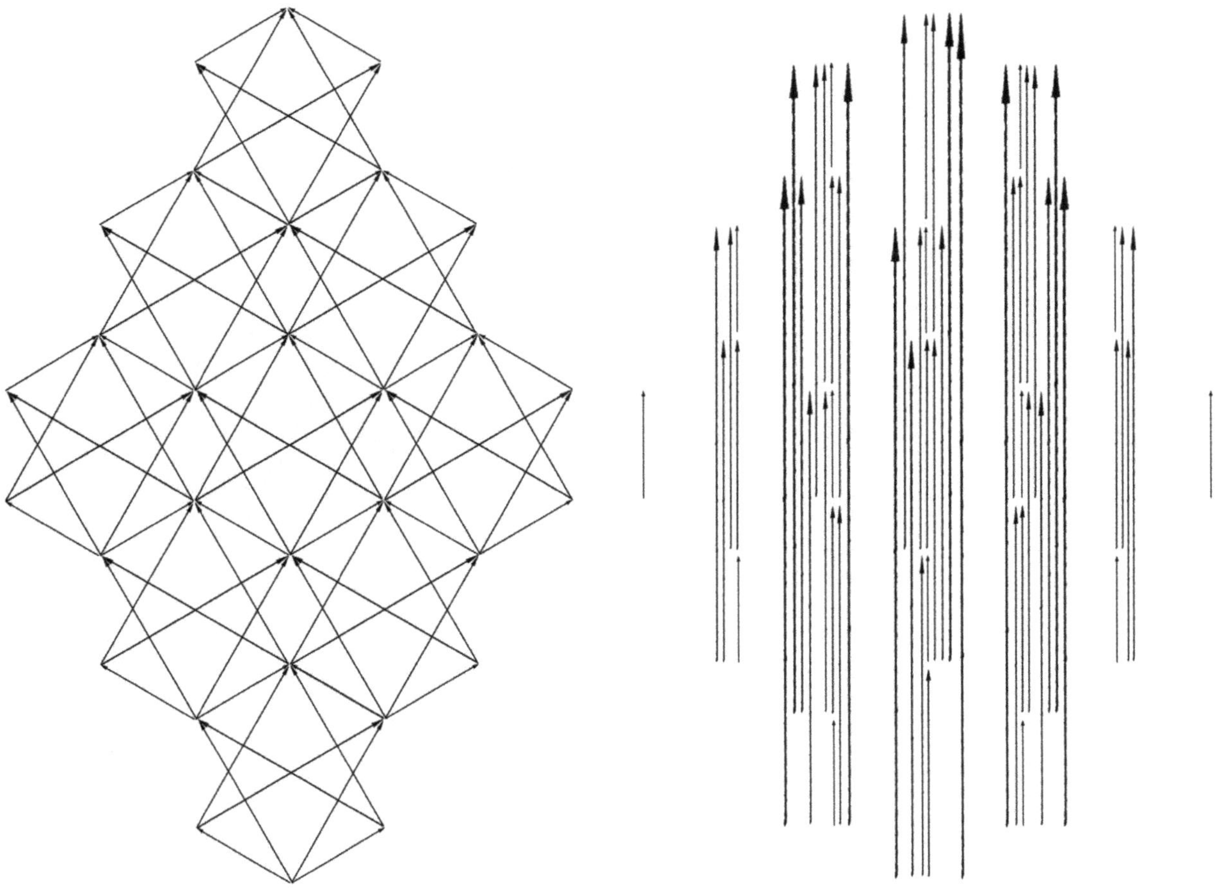

I suggest that the above causal set diagram provides the interpretation for the fine-structure constant (FSC). The "hexagon tiling scheme" is just barely established in the case of N = 3. Honeycombs of higher index than N = 3 consist of either a hydrogen cycle (N is even) or a helium cycle (N is odd) enclosed by concentric rings of hex cells. Thus the cycle diagrammed above is a very basic substructure of any more extensive four-dimensional latticework, providing a basis for the calculation of charge-quanta occupancy arrangements in space-time, namely quantum electrodynamics.

As N becomes large, the number of proper-time quanta locations approaches $(4/3)N^3$. The limiting density of quanta count per unit volume is obtained by dividing $(4/3)N^3$ by the four-dimensional volume of the region in hex cycles, N^2. So the capacity limit of the honeycomb series for proper-time quanta per unit volume approaches (4/3)N. This may have application in cosmology. The mass-energy calculation for a given region of space-time that is mostly vacant of atomic nuclei concerns the four-dimensional lattice quanta together with a complement of proper-time quanta. Here we have a basis for calculating the mass of "empty space," which should explain the "dark energy" discrepancy in cosmological accounting.

Quantum Mechanics

Quantum mechanics employs a four-tuple of integer values to specify the "electron state" of an atom. The four integer parameters are associated with four concepts tied to Newtonian physics: orbit number (Bohr orbit), sub-orbit number, angular momentum, and up/down spin. If we set out to populate our cloud formations with quanta of charge and momentum, we foresee a limited range of available "fill patterns." A lone hex cycle offers two "slots" for charge quanta and three more slots for quanta of forward momentum, as shown previously. The number of fill patterns grows systematically with increasing cloud size. The concentric rings of hex cycles, in this theory of cloud structure, provide the orbit and sub-orbit numbers for the four-tupled values of QM. The values associated with angular momentum must indicate alternate fill patterns that can align in off-axis orientations to the proper-time axis. The up/down spin parameter doubles the number of "electron states" that are due to the first three parameters.

Looking back at the explanation for spin that was proposed earlier, it seems that the spin orientation of any one hex cycle in a cloud must match the spin value of all hex cycles in that same cloud, yielding only two alternatives for each possible cloud-seeding arrangement that would otherwise obtain without taking spin into account. This would explain the doubling effect of the spin parameter on the total number of possible cloud-fill patterns.

I have put the term "electron state" in quotation marks because there is no such thing as a state in this theory. Sheer temporal succession is purely dynamic, providing no such thing as an instantaneous state. A quantum is certainly not a state. The only static entity is the generic individual moment, which has no variability or numeric value; thus the "state" of a moment has no utility in physics. Instead of states, there are distinct arrangements of quanta that characterize the variety of pathways from the earliest moment to the latest moment of any bounded region of time. A given cloud cycle is such a bounded region of time, and the four-tupled values of QM characterize the possible arrangements of quanta in that cloud cycle.

Nuclear Structure, the 6-D Time Lattice

Attached to periodic nodes of an electron cloud sequence is a nuclear sequence comprised of higher-frequency quanta. This constitutes a synchronization of the nucleus to its electron cloud and assigns the nucleus its location in the four-dimensional manifold. The "attachment" of nucleus to cloud can only mean that the two distinguishable sequences share periodic nodes. In general, "attachment," "connection," "contact," or "collision" between distinguishable propagation sequences can only mean that such sequences develop shared nodes in common.

The following photograph of three ball-and-stick models illustrates the construction principle of the nuclear time lattice. This lattice will supplant the concept of a nucleus composed of quarks and gluons. The magnetic struts cling to the steel ball bearings. All struts are the same length. The color of a strut has no significance.

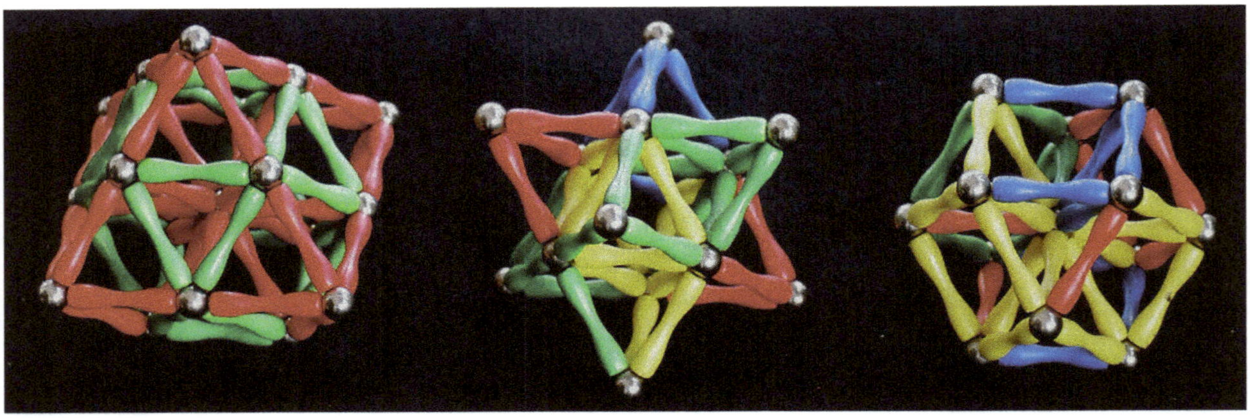

The three assemblies shown are substructures of a single lattice. The assembly in the middle can be described as a unit octahedron with a tetrahedron erected on each face. On the right we see a cuboctahedron, with six square faces and eight triangle faces. Caged inside the cuboctahedron is a center ball bearing with twelve radial struts connecting to the twelve outer vertices. If we erect a pyramid on each square face of the cuboctahedron, we get an octahedron of edge length two, as shown on the left. (Its bottom pyramid is omitted in order to allow the assembly to stand upright.)

The principle of the lattice is this: each unit tetrahedron shares its faces with octahedra, and vice versa; no tetrahedron or octahedron shares a face with one of its own kind. Following this rule, we can fill space with a mix of unit tetrahedra and unit octahedra. As we fill space, the above forms will recur periodically in the lattice. Larger versions will also be formed, at twice the size, three times the size, etc. This lattice principle has widespread practical use in building design, for which Buckminster Fuller named it "the octet truss."

The spatial lattice exemplified by the models becomes a time lattice when each strut is assigned an unambiguous direction while obeying chronology protection. Each strut is then a quantum. We thereby obtain a richer structure, a causal set, composed entirely of time-directed pathways.

We are exploiting well-known regular solids of high symmetry in order to obtain causal sets that exhibit comparably high symmetry. When converting a finite spatial topology to a causal set, the directionality can be assigned in such a way as to maximize the preservation of symmetry. The "time tracing" can yield a causal set with symmetry that rivals the topology of the regular solid being traced. In tracing the above models in order to arrive at causal sets, there are four axes of orientation that yield high symmetries. We can specify each of these axes by reference to the cuboctahedron model:

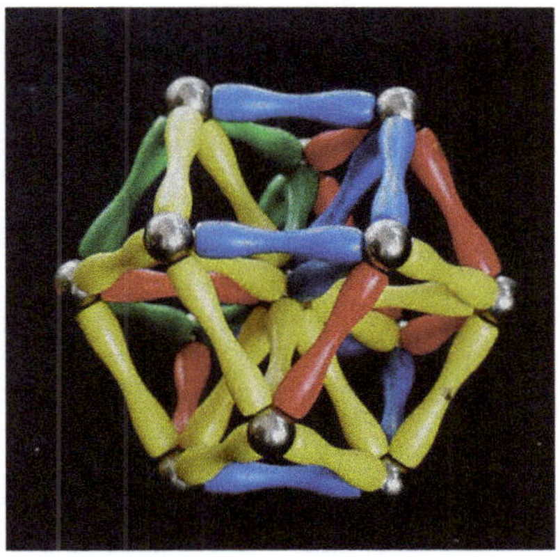

One axis runs from the center of a triangle face to the center of the diametrically opposed triangle face. This results in three-way symmetries about a proper-time axis. An alternate axis runs from the center of a square face to the center of the opposing square face, resulting in four-way symmetries about that time axis. A third axis runs from any vertex of the cuboctahedron to the diametrically opposed vertex. Finally, an axis of high symmetry runs from the center of any edge of the cuboctahedron to the center of the edge that is diametrically opposite.

The cuboctahedron model has one interior node joining twelve struts. These twelve struts become twelve causal links once we've performed a time tracing. There will be six quanta arriving and six departing at the interior node. As the lattice is extended, every interior point will have that same context. The nuclear lattice structure is thus a six-dimensional time lattice. We now have the terms "four-dimensional lattice" and "six-dimensional lattice" to distinguish electromagnetic structure from nuclear structure. The four-dimensional lattice is the more extensive of the two, since it is space-time. The six-dimensional lattice only extends far enough to form highly symmetrical structures that can propagate on their own by chained repetition. These nucleons of characteristic six-dimensional structure have their location in the more extensive four-dimensional lattice by sharing periodic nodes with their companion electron clouds.

Structure of the Proton

The proton has the overall structure of a cube, for which the model pictured above can be referenced. The eight outermost ball bearings define a three-dimensional cube that envelops the discrete structure. The model shows the minimal cube that can be formed using the six-dimensional lattice principle. It's actually an eight-pointed star, but I'll call it a "small cube." The model is resting on one of its sides, but the camera angle almost provides a line of sight through the small cube from one corner to the opposite corner. This is the time-tracing axis by which the spatial topology is converted to the causal set structure of a proton. Such a tracing yields three-way symmetries among the constituent quanta about the axis of proper time. Quanta that break the three-way symmetry are color-charge quanta. To form a stable proton, the color-charge quanta must form three-way symmetries amongst themselves in a suitable arrangement.

A larger cube can be constructed by assembling eight small cubes together in a 2 × 2 × 2 arrangement. The next larger cube after that is twenty-seven small cubes in a 3 × 3 × 3 assembly. Here I believe we have arrived at the structure of a single proton. The small cube has thirty-six struts. Thus the 3 × 3 × 3 assembly has roughly (27 × 36) struts. 1836 = (27 × 34) × 2, which is the empirical mass ratio of proton to electron. The approximation of (27 × 36) to (27 × 34) is a clue that the 3 × 3 × 3 assemblage of small cubes will yield the causal set structure of a single proton cycle.

Inspecting the small cube model along the line of sight, just below the nearest vertex, one sees a triangle of three yellow struts that is perpendicular to the time-tracing axis. There is a second such triangle further along that axis, diametrically opposed. These two triangles contribute to the three-way symmetries of the spatial form. As time triangles, however, they present a symmetry problem. Three radial directions are possible for one of the triangles, and three more for the other triangle, for a total of six distinct radial directions. Six of these "axis-hugging" triangles are needed to achieve a radial symmetry in the proton cycle. A 3 × 3 × 3 assembly of small cube cycles has three small cubes stacked up in chained repetition along the major time axis. Those three cycles contain two problematic triangles apiece, for a total of six. Thus the six triangles can assume the six different radial directions, salvaging radial symmetry for the overall proton cycle.

Structure of the Neutron

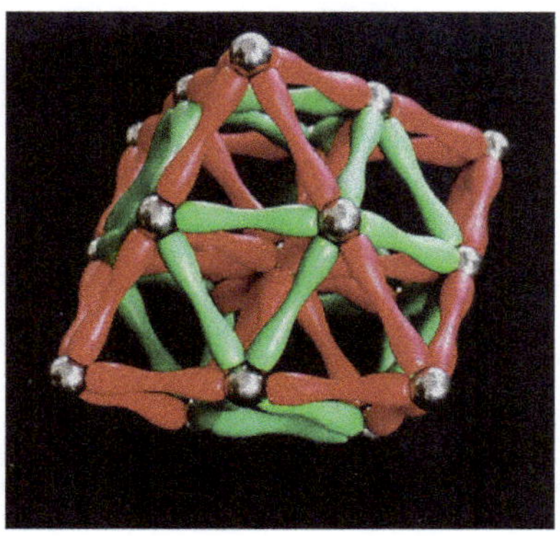

If we view this model from directly above, we see nothing but four-way symmetries, which suggests a neutral orientation with respect to the four-dimensional electron-neutrino lattice. The octahedron has eight faces and six vertices, while the cube has six faces and eight vertices. The two Platonic solids are called "duals" for that reason. When constructing models on the six-dimensional lattice principle, the cube and the octahedron serve to compound one another in various ways. This will account for the role of protons and neutrons as major substructures of the nucleus. At this stage of development, one must conclude that the simplest, most symmetrical topologies correspond to the most primordial substructures of matter. Both the proton and neutron have their quanta in the six-dimensional lattice arrangement. A proton has the overall topology of a cube. A neutron has the overall topology of an octahedron. The above model is an octahedron of edge length two. A larger version of the octahedron, of edge length six, will yield a model of the neutron, as counterpart to the 3-cube model of the proton.

Mass Values of Proton and Neutron

The calculation of energy ratios is not trivial in general. Each quantum must be weighted by its relative frequency, which involves the whole network of its neighborhood within some specified bounded region. However, in the case of a regular lattice, if the lattice quanta are of uniform frequency, energy calculations can be as simple as counting quanta. The first step in calculating the mass-energy of the proton is to count the lattice quanta in the 3 × 3 × 3 assembly of small cubes that makes up a proton cycle.

My formula for the number of struts in an N × N × N assembly of small cubes is:

N-cube struts = $24N^3 + 12N^2$

By the formula, the 1-cube model (the "small cube") calculates to thirty-six struts, which is correct. The 2-cube has 240 struts, and the 3-cube has 756 lattice struts.

The 3-cube lattice assembly provides locations for additional quanta at stepped-down frequency values, just as the four-dimensional lattice provides locations for its own lower-frequency quanta. For instance, each "X" on the face of a small cube is made up of four struts. A half-frequency strut, twice the length, can span each small-face diagonal also, adding 108 struts at half frequency to the 3-cube. This would add 54 to the 756 energy calculation.

My formula for the number of lattice struts in the octahedron of edge length N is:

Octa-N struts = 12 × S(N)

(S(N) is the sum of squares from 1 to N.)

The unit octahedron (N = 1) calculates to 12. The Octa-2 (the model pictured) calculates to 60 struts. The Octa-3 has 168 struts. The Octa-6 has 1092 struts. The Octa-6, with an edge length of six struts, is the finite structure proposed for a neutron.

As oriented in the photo, an octahedron has a number of "horizontal floors" that consist of square tiling arrays. The time tracing of these horizontal struts presents symmetry problems, so some of those struts might be excluded from the neutron, reducing the count from 1092. The "odd-numbered floors" are particularly difficult to time trace with any great symmetry. These are the floors with 1, 9, and 25 tiled squares. By omitting these floor struts, we reduce the Octa-6 strut count by 176, to 916.

To summarize, we have the 3-cube for a proton and the Octa-6 for a neutron. Each face of the 3-cube has diagonals that are six struts in length, while the Octa-6 has an edge length of six struts. The 3-cube has three small cubes in chained repetition forming its core along the major axis. The Octa-6 has six unit octahedra in chained repetition that form its core. Lattice count for the two assemblies is 756 and 1092, respectively. The sum of those two numbers is 1848, and the average is 924. The lattice count of the 3-cube is 168 short of the average, and the lattice count of the Octa-6 is 168 high of the average. (168 happens to be the strut count of an Octa-3.)

We have lowered the neutron mass close to the 924 average by deleting some floor struts. We've added weight (+54) to the proton model by including some half-frequency quanta. Another 108, or 27 × 4, is needed to bring the proton mass up to 918 (half of 1836). We have not yet considered any locations in the 3-cube, or any locations in the Octa-6, for quanta in the proper-time axis. I will assume at this point that further detailed considerations can bring the modeling of the proton cycle and the neutron cycle to their empirical mass values, or rather to their half-values.

Next we consider the compounding of protons and neutrons to make larger nuclei, and the relation of a nucleus to its electron cloud.

The Integration of Nucleus and Electron Cloud

A beautiful thing about the cuboctahedron is that its edges are formed of four regular hexagons, which encircle the solid like great-circle routes. In fact, the octet truss can be assembled entirely from regular hexagons. The hex cycles of the four-dimensional lattice are also hexagonal in their topology.

We have already taken some account of lower-frequency quanta, lower frequency than the lattice struts, that can populate regions of either lattice type. Now consider the Octa-6 formation. It has the overall contour of a single octahedron with six vertices. These six vertices are connectable by twelve quanta to form a low-frequency octahedron that borders the whole assembly. A ladder of discrete nuclear frequencies extends from the high-frequency value of the six-dimensional lattice struts to the lowest-frequency quanta of the "envelope octahedron" just described. At this point, the lowest nuclear frequencies match up to the frequencies of the hex cycles of the four-dimensional lattice.

The four hexagons that encircle a cuboctahedron are also present on the surface of an octahedron or a cube. The four hex cycles of a hydrogen cloud might therefore be situated "on the surface" of a proton, having assumed the identity of the low-frequency cuboctahedron that the proton structure accommodates. The nuclei of atoms in the Periodic Table might variously exhibit the overall form of a cube, an octahedron, a cuboctahedron, or compounds and stellations of those forms. Whatever form the larger nuclei might take, they will afford low-frequency hexagon locations to accommodate the low frequencies of time-dilated hex cycles.

Since the hexagons of the six-dimensional lattice assume four distinct orientations, only one of the four orientations can mesh with the four-dimensional lattice. Only one orientation of a nucleus can be brought into line with the proper-time axis of the local four-dimensional manifold. A given form of nucleus will have one of its low-frequency hexagons properly oriented to mesh with the neighboring hex cycles of its electron cloud, fitting the nucleus into a honeycomb cycle.

In this paper, rudimentary forms have been used to model the nonperturbative persistence of stable particles. Collision and interaction of particles involve transitional forms with short persistence. These intermediaries are presumably deficient in respect to their inherent symmetries. The detailed topology of perturbative sequences is a task well suited to experts in the field of regular polygons, and more generally the field of regular polytopes. These lend themselves readily to conversion to causal sets, at which point energy calculations can be brought to bear.

Summary

This theory offers a solution as to what particles are, what mass is, and why particles have the masses they do. Particles are repetitious patterns of time sequence, and the mass-energy of a particle is the measure of its inherent relative frequencies, which is a function of the number and arrangement of its constituent quanta. A particle sequence as a whole must in turn have a frequency relative to other particles, since all frequency is relative. All quanta of this universe connect into one structure. That is what "this universe" means. The causal connectivity is what rations the frequencies, or energies, of this universe.

I came to these constructions via Russell and Whitehead's doctrine of space-time called "eventism." They had developed the event ontology in response to Special Relativity. The ontology they embraced is one of momentary monads, akin to Leibniz's monads. The epistemology and phenomenology that underlies causal set theory was all worked out by Russell in Human Knowledge: Its Scope and Limits. The phenomenology of mind, as for example the sensuous visual field, is a domain of true spatial relations and geometry. However, the subject matter of physics is a world external to our perception, purely sequential in nature, and devoid of instantaneous spatial relations.

As an isolated researcher, I refer to my own publications for further background on the reduction to time. The diagrams, and the recognition of frequency ratios as energy ratios, first appear in The Mind-Body Problem and Its Solution, 2004. Subsequent development of the diagrams, to the point of Bohr's formula, appear in A Theory of Everything for Physics, 2005. An article, "Finite Eventism," appears in the book collection Mind *That* Abides: *Panpsychism* in the *New Millennium*, 2009, Benjamins Publishing. The article covers much the same ground as this paper but adds a section on the functional role of the cortical homunculi, under the assumption that the brain is a causal set.

A FREQUENCY RATIO ACCOUNT OF TEMPORAL ATOMISM

Author(s): Carey R. Carlson

Source: Process Studies, Spring–Summer 2021, Vol. 50, No. 1 (Spring–Summer 2021), pp. 107–127

Published by: University of Illinois Press

Stable URL: https://www.jstor.org/stable/10.5406/processstudies.50.1.0107

A Frequency Ratio Account of Temporal Atomism

Carey R. Carlson

CAREY R. CARLSON is an independent researcher who studied the philosophy of science under Grover Maxwell at the University of Minnesota. His recent book is The Mind-Body Problem and Its Solution (2nd ed.).

Abstract

This article examines the time duration of individual occasions in the light of the discovery that temporal succession produces frequency ratios. The frequency ratios are used to define energy ratios and the quantum. The manifold and the common particles are constructed graphically using the arrows of time, with the mass ratios of the particles derivable from the graphs. The formal reduction of physics to time compels us to adopt Whitehead's conception of the physical universe as occasions of experience engaged in temporal and causal succession. The relative duration of the constituent occasions of the particles is determined by their graphs. In the final section, a refined account of the mind-brain interaction sequence confirms the duration of a human occasion as one-tenth of a second.

Introduction

Alfred North Whitehead's Science and the Modern World provides a clue to the hard problem of mind and body. A sensory quality such as redness is no part of mathematical physics. The theoretical entities of physics owe their formulation to conjecture, while entities like redness belong to the immaterial, dreamlike experience of sentient mind. Hence, the bifurcation of nature into "the conjecture and the dream." A solution is intimated in Russell's Human Knowledge. Russell contends that the physical world, including the four-dimensional manifold, is unperceivable, and that physics can at best discover the "causal skeleton" of the world, based on systematic conjecture as to the external causes of our phenomenal percepts.

The brain is in the head, but thoughts are not, so, at least, philosophers assure us. This point of view is due to a confusion between different meanings of the word "space." Among the things that I see at a given moment there are spatial relations which are a part of my percepts; if percepts are "mental," as I should contend, then spatial relations which are ingredients of percepts are also "mental." Naïve realism identifies my percepts with physical things; it assumes that the sun of the astronomers is what I see. This involves identifying the spatial relations of my percepts with those of physical things. Many people retain this aspect of naïve realism although they have rejected all the rest. But this identification is indefensible. The spatial relations of physics hold between electrons, protons, neutrons, etc., which we do not perceive; the spatial relations of visual percepts hold between things that we do perceive, and in the last analysis between colored patches. (Russell 201–202)

When one realizes that Descartes's extension in space is unperceivable, one may grasp the idea of reducing dual-parameter space-time to a causal web of time-ordered events. That is a purely conceptual readjustment on the conjecture side of "the conjecture and the dream." Familiarity with Whitehead leads one to conclude that the role of sentient mind in the physical world is resolved by filling the causal web of events with Whitehead's occasions of experience.

For example, let the working hypothesis be that the ultimate realities are the events in their process of origination. Then each event, viewed in its separate individuality, is a passage between two ideal termini, namely, its components in their ideal disjunctive diversity passing into these same components in their concrete togetherness. There are two current doctrines as to this process. One is that of the external Creator, eliciting this final togetherness out of nothing. The other doctrine is that it is a metaphysical principle belonging to the nature of things, that there is nothing in the Universe other than instances of this passage and components of these instances. Let the latter doctrine be adopted. (AI 235–236)

Grover Maxwell shared my understanding, although he was not familiar with Whitehead. The present article is an attempt to add to Maxwell's view by claiming that the arrows of time form frequency ratios. I will show that the manifold and the common particles are produced by the simplest repetitive patterns that time can make.

I used "causal set theory" in the title of the above quotation so that physicists could relate my findings to an already established branch of physics. Causal set theory formalizes the discrete event ontology. The theory employs directed graphs to specify formations generated by a temporal and causal successor relation. Such formations are called "causal sets." The moments and transitions of time are termed "elements" and "causal links." Rafael Sorkin, who founded causal set theory, initially thought he had found the theory of everything. He asserted that causal sets satisfy special relativity "out of the box," that causal sets are just time, and that the world is a causal set. Before I ever heard of causal sets by name, I had employed the equivalent of causal sets and causal set graphs in The Mind-Body Problem and Its Solution. For documentation of my two main sources, I selected passages from Russell's Human Knowledge and Whitehead's Adventures of Ideas.

Einstein used the limiting velocity as an unexplained postulate to derive the rest of special relativity. There are fresh TED Talks online with physicists asserting that there is still no explanation for the limiting velocity. Causal sets explain the limiting velocity because any change of spacelike interval is a purely structural consequence of the stepping action of time. Thus, space cannot outrun time. The continuing pursuit of theories that do not explain the limiting velocity indicates the persistence of belief in instantaneous extension in space.

I am not a Whitehead scholar, and I am highly selective regarding Whitehead's ideas that I employ. However, his panpsychism was the incentive that led me to graphs of time and the frequency ratios formed. Once found, this discovery stands on its own merits without the support of Whitehead's interpretation. The discoveries are offered here for their effects on Whitehead scholarship, not the other way around. I will simply introduce the frequency ratios and follow the logical consequences in accord with the following prescription.

In framing a philosophic scheme, each metaphysical notion should be given the widest extension of which it seems capable. It is only in this way that the true adjustment of ideas can be explored. More important even than Occam's doctrine of parsimony, if it be not another aspect of the same, is this doctrine that the scope of a metaphysical principle should not be limited otherwise than by the necessity of its meaning. (AI 237)

Temporal Succession and Its Frequency Ratios

Was Whitehead a temporal atomist? The following passage strongly suggests that he was.

The notion of the contiguity of occasions is important. Two occasions which are not contemporary are contiguous in time when there is no occasion which is antecedent to one of them and subsequent to the other. A purely temporal nexus of occasions is continuous when, with the exception of the earliest and the latest occasions, each occasion is contiguous with an earlier occasion and a later occasion. The nexus will then form an unbroken thread in temporal or serial order. The first and the last occasions of the thread will, of course, only enjoy a one-sided contiguity with the thread. (AI 202)

In a continuous series, there is always another moment between any two, so with respect to any given moment there is no next moment. Thus, no two moments are contiguous in a continuous series. Since Whitehead stresses that "the notion of the contiguity of occasions is important," I believe that Whitehead was a temporal atomist. It is unlikely he would be careless with the word "contiguity."

A discrete time series can be illustrated by a directed graph of arrows connected tail to head (see figure 1). Each node represents a moment of time, and each arrow represents a discrete temporal transition from one moment to another.

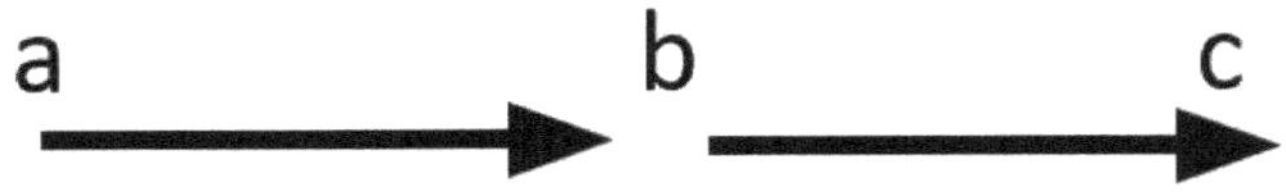

figure 1 *A Discrete Time Series.*

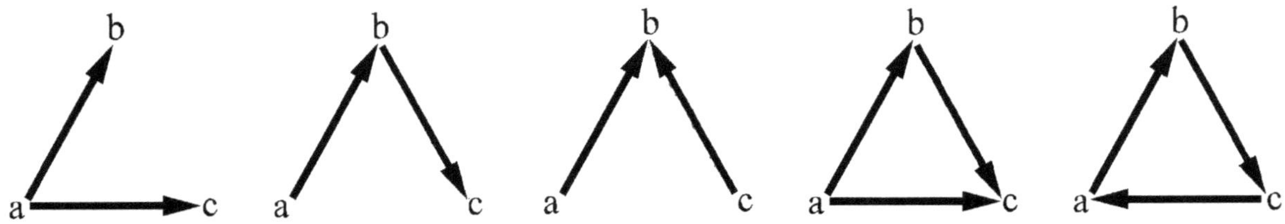

figure 2 *Time Graphs with Exactly Three Moments.*

In figure 2, the nodes of the diagram (the moments) can represent occasions of experience. The connecting arrows (the temporal transitions) then represent Whitehead's physical prehensions. Temporal succession is conflated with causal succession, so that a moment of time can have more than one successor and more than one predecessor. There are four ways that three moments can be connected by time steps. The graph on the far right of figure 2 is not a time graph, because each occasion is its own causal ancestor and its own causal descendant. No occasion can come before or after itself.

From left to right we see a fork in time, then a time series of two steps, then a convergence of time steps, and finally a time triangle that features forking and convergence in combination to form a reconvergent sequence. The time triangle is the simplest graph to feature frequency ratios. It has a two-step path from a to b to c and a one-step path from a directly to c. Both paths begin at the same time and end at the same time. Thus, both paths span a single common time interval, which affords comparison of the frequencies of the two paths. One path takes two steps while the other path takes one step to traverse the common time interval, forming a frequency ratio of 2:1. Figure 3 is another graph that shows the frequency ratio 2:1.

There are three causal routes from the bottom event to the top one. Either side route involves two causal transitions with an intermediate event along the way. All three routes determine the same temporal interval, since they all begin at the same time and end at the same time. Either side route has twice the temporal frequency of the middle route.

Figure 4's alternate routes between common end points have two arrows forming one route and three arrows forming another, producing a frequency ratio of 2:3. Time diagrams exist for frequency ratios ranging across the rational numbers.

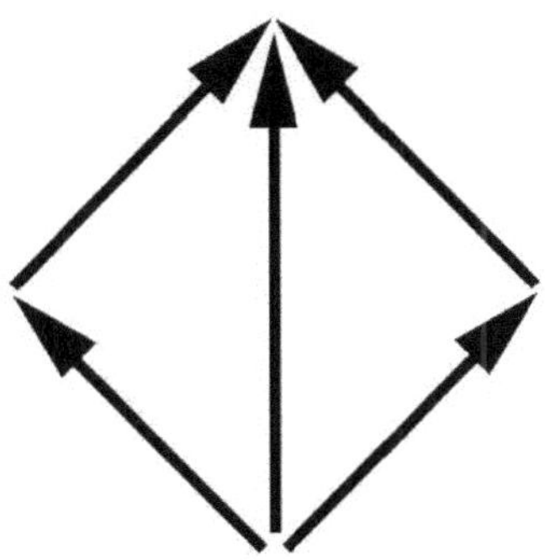

Figure 3. Frequency Ratio of 2:1.

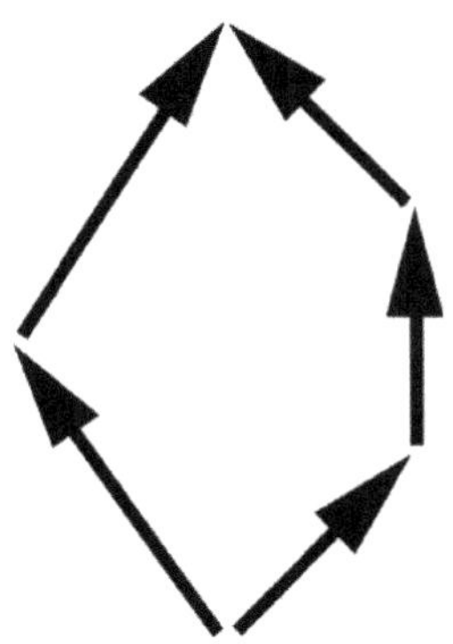

Figure 4. Frequency Ratio of 2:3.

Defining Energy in Terms of Frequency

Planck discovered the coupling of energy and frequency at the quantum level, expressed in the formula $E = hf$. Thus, two energy values, E_1 and E_2, are equal to two frequency values, hf_1 and hf_2. The ratio of two energy values, E_1 / E_2, is equal to the ratio of two frequency values, f_1 / f_2. Planck's constant drops out. The frequency ratios can define the energy ratios. The step of discrete time is the unit of the frequency ratios. Thus, the step of time is the quantum of the energy ratios. We have a structural definition of energy and its quantum in terms of time alone, illustrated in the simplest case by the time-triangle diagram of three arrows. Directed graphs can now serve as quantum schematics, each arrow depicting a quantum.

Notice that the physical prehension, not the occasion, is the unit of the frequency ratios and the quantum of the energy ratios.

Relative Duration

Doing away altogether with the dimensional units of physics, the metric of temporal succession will be based entirely on frequency ratios as shown by arrow diagrams. Only ratios will have absolute values. Empirical measurement always involves comparisons. Even the second of time is an arbitrary unit of duration. That is why a ratio-based metric is ideal for the theoretical foundation of physics.

The reciprocal of a frequency value is a time period or measure of duration. A quantum inherits the relative frequency value of the sequence to which it belongs. The reciprocal of a quantum's relative frequency is its relative duration. Higher-frequency sequences consist of quanta with higher frequency and shorter duration than the quanta of lower-frequency sequences. The physical prehension, or temporal transition, is the unit of frequency ratios. Thus, the prehension, which is the quantum, has the duration value. Judging from the arrow diagrams, the occasion has no duration of its own. We must include prehensions in order to gauge the lifetime of an occasion. Since an occasion "has its being" in its becoming and perishing, it makes sense to include its physical prehensions in the accounting of its duration.

Space-Time Replaced by a Four-Dimensional Time Manifold

The usual understanding of physical space is confined to geometric features, such as areas, volumes, points, and lines. These features can be ascribed to a person's visual experience, even during dreaming. Science must have a space that is consistent for all observers, a harmonization of perspectives based on waking perception, discarding the data of imagination and dreams.

Furthermore, the space of physics must do without color or sensory qualities in its definition. It follows that we do not perceive physical space. We are restricted to conceiving it. The same holds true for the entities that occupy physical space. The whole apparatus of physics is refined from commonsense belief in a world not limited to our sensory experiences.

Though we might be aware of the distinction between perceived space and the space of physical theory, we unhesitatingly borrow the geometric features of visual space and carry them over to the space of physics, as we do when learning Euclidean geometry in school. We fail to distinguish pure geometry, which is mathematics, from the geometry of our actual world, which is physics. The requisite distinction is found in the very purpose of science, which is to build a predictive causal framework for our scattered perceptions.

Einstein remarked that a discrete manifold could provide an inherent metric, but he never foresaw the discrete time manifold, much less the inherent frequency ratios it provides. The concept of a manifold created by sheer temporal succession is clarified by diagrams in which the arrow of time is the only graphic element. Figure 5 shows an extendable discrete time manifold of two dimensions. It serves to illustrate Whitehead's definition of contemporary events: "It is the definition of contemporary events that they happen in causal independence of each other. Thus two contemporary occasions are such that neither belongs to the past of the other. The two occasions are not in any direct relation to efficient causation" (AI 195). Occasion "E" in figure 5 has its own unique set of past and future occasions. Occasions within the unshaded regions are the contemporaries of E. A contemporary of E is causally connected to E only indirectly via common causal ancestors or common causal descendants. Two contemporaries are "causal cousins." Neither is in the direct lineage of the other.

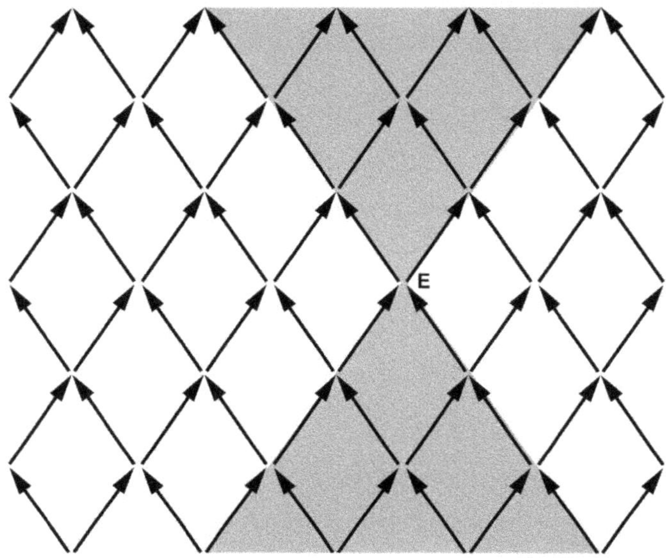

Figure 5. The Causal Past and Causal Future of E.

With Newtonian space, an instantaneous "now" sliced across the spatially extended universe. With special relativity, the "now" is proprietary to each occasion. The discrete time manifold illustrated in figure 5 has no primitive spatial relations to support instantaneous spatial extension. Temporal transitions alone suffice to form the manifold. In the passage quoted above, Whitehead refers to a "purely temporal nexus of occasions" when referring to a strictly serial nexus. Figure 5 depicts a nexus that is not purely serial, yet it is purely temporal.

Notice also in figure 5 that if occasion E did not exist, then the four transitions that connect E to other occasions would not exist either. That would form a hole in the manifold, which is a legitimate alternative possibility. Attributing perfect uniformity to the manifold restricts the structural possibilities of temporal succession, ruling out options that are important for the reduction of physics to time.

The "hex cell," consisting of six moments connected by ten transitions, can be used to construct an extendable four-dimensional time manifold (see figure 6).

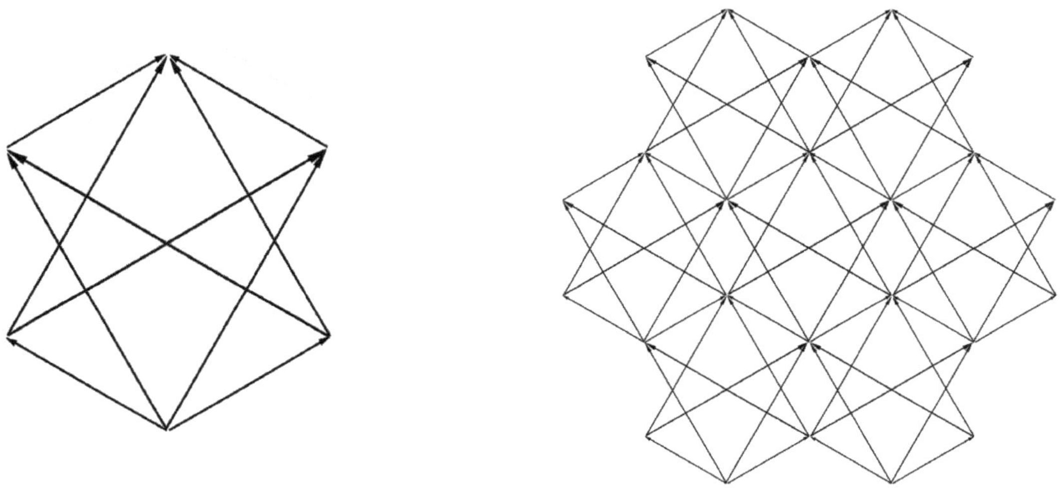

Figure 6. An Extendable 4-D Time Manifold.

Each interior node has four arrows arriving and four departing, the defining characteristic of four-dimensionality. The diagram can be extended with more hex cells to depict four-dimensional regions of any extent. Such a four-dimensional region is made of quanta, like everything else in the theory of sheer temporal succession.

A "bounded region" can be defined as a region graphed by a "closed diagram," a diagram that has a single earliest moment and a single latest moment. Any closed diagram can be repeated as a cycle in chained repetition, such that the last moment of one cycle is the first moment of the next. By building up physics from closed diagrams, we confine ourselves to what can be confirmed by experiment. Any experiment is bounded by two moments, the triggering moment when the experiment is initiated and the final moment when the outcome is known. In between, nature enacts a causal sequence that results either in the predicted outcome or something else.

The Reduction of Particles to Time

The hex cell that was used to make a four-dimensional manifold will also serve, in chained repetition, to make an electron. The hex cell has locations for additional quanta that provide charge, momentum, and spin (see figures 7 and 8).

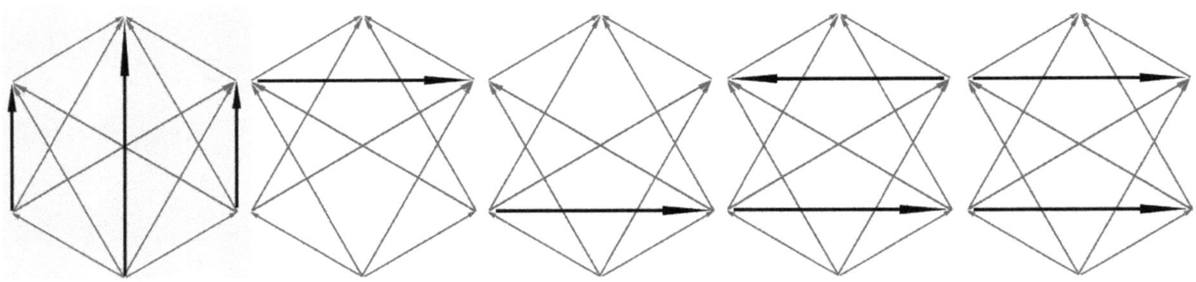

Figure 7. Quanta of Charge and Momentum.

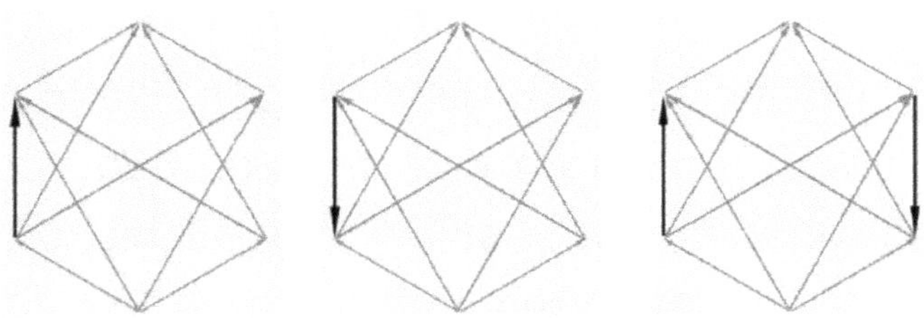

Figure 8. Spin Features.

The vertical arrows represent quanta of forward momentum. The horizontal arrows, which necessarily break the bilateral symmetry of the cell, represent charge quanta. Spin options give the cell a type of "handedness." (The hex cell accommodates one "down arrow" at the most. Two down arrows would form a closed loop, violating chronology protection.)

A bare hex cell with no charge quanta may be called a "neutrino cycle," because it forms, in chained repetition, a neutrino. A hex cell with charge quanta may be called an "electron cycle," because it forms, in chained repetition, an electron.

Bohr's Formula

The following set of four sequences in figure 9 show a free electron, electron clouds of hydrogen and helium, and a cloud that absorbs and reemits a photon. The whole atomic series follows suit with these initial sequences in producing the discrete Bohr line frequencies for photon absorption and emission by the atoms.

I have drawn the hex cycles as neutrino cycles, without any charge quanta. As drawn, the sequences depict modes of neutrino propagation. If we populate the hex cycles with charge quanta, we get electron clouds. In that case, we would have, from left to right, a free electron, a hydrogen cloud, a helium cloud, and finally a hydrogen cloud sequence disturbed by an encounter with a photon. The first three diagrams mark the start of a progression that continues in step with the periodic table.

The stable clouds feature uniform cycles of uniform frequency. The f-number above a stable sequence is the frequency of its cloud-cycle repetition. The cloud disturbed by the photon exhibits a modulation of cycle pattern and frequency. The frequency of the photon, either incident or emitted, is the difference in frequency of the two cycles involved in the modulation. This scenario, generalized to the whole series of cloud possibilities, yields Bohr's formula for the frequency and wavelength values of photons absorbed and emitted by the entire atomic series.

A free electron is an electron cycle in chained repetition. Its frequency of repetition is f_1 as labeled in figure 9. Taking f_1 as the provisional unit of frequency, the wavelength of each cycle is also 1. Each sequence is scaled so that thirty-six hex cycles span the same height on the page, indicating that the four sequences elapse in the same amount of time. From the diagram, the wavelength of a hydrogen cloud cycle is four times that of the wavelength of a free electron cycle. The wavelength of a helium cloud cycle is nine times that of a free electron cycle. The wavelengths 1, 4, and 9 correspond to frequencies 1, 1/4, and 1/9, respectively. The intruding photon has the frequency $f_2 - f_3$, or 1/4 – 1/9. Bohr's formula for frequency, neglecting a constant scale factor, is $1/k^2 - 1/n^2$. For integer values k = 2 and n = 3, our graphical account gives the same photon frequency as Bohr's formula.

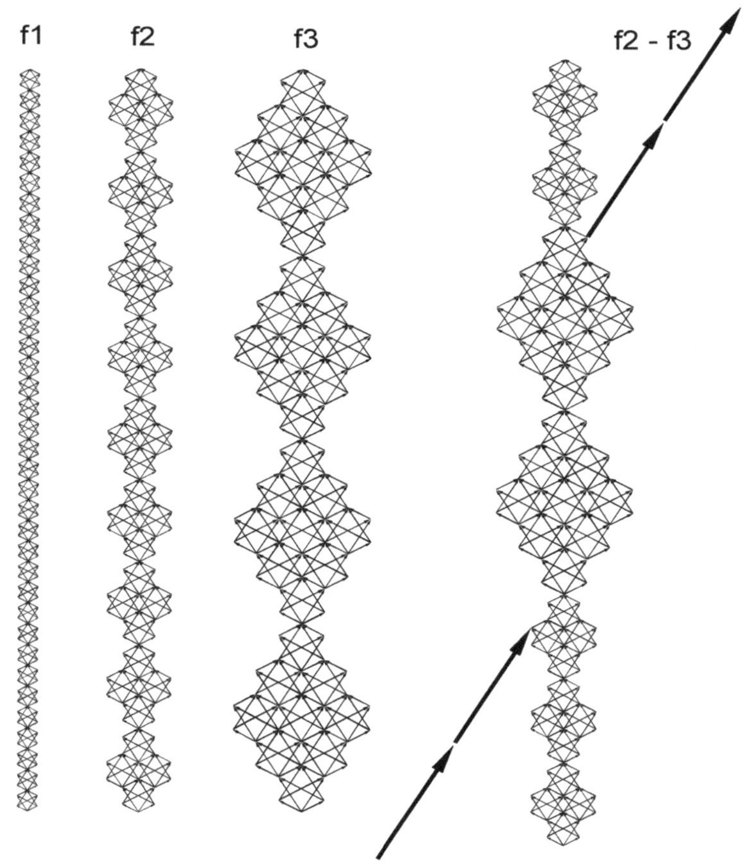

Figure 9. Graphical Account of Bohr's Formula.

Notice that the electron as graphed has frequency and wavelength values. The wave–particle duality is resolved by means of discrete constructions. Even the electron clouds have their own frequency and wavelength values.

If we take the electron cycle as the unit of energy, then the progressive scaling that stretches thirty-six cycles to the same height on the page for each sequence means that the same amount of energy plays out in the same amount of time for each sequence. Thus, a type of symmetry is the basis of the discrete time dilation from one cloud sequence to the next. The discrete time dilation can serve as the basis for a discrete version of general relativity.

General relativity uses the curvature of space-time to describe gravity, with time dilation being a consequence. An ad hoc postulate states that the amount of curvature is proportional to the local mass density. Assuming the nuclei dilate in step with their electron clouds, greater nuclear mass is tied to greater time dilation, explaining gravity with a symmetry principle rather than an ad hoc postulate.

Bohr's formula encapsulates a wealth of spectroscopy data. An explanatory model of the underlying structure of electrons and their behavior should not be more complex than the empirical data it is supposed to explain. A model represents theory, while Bohr's tidy formula represents the empirical data. We now have an account of electron–photon dynamics that is nearly as simple as the formula that encapsulates the data. We have no fundamental units of measure. We are left with the diagrams themselves, in which the electron appears as the real-time clock that sets the base frequency for the electromagnetic spectrum.

It is evident that four-dimensionality arises at the smallest scale with the electron and neutrino. Electron clouds and neutrino formations share the telltale hex-cycle structure of the four-dimensional manifold, so we may speculate that they constitute what there is of the manifold. Gaps in the manifold delineate the locally propagating sequences of neutrino and electron formations. That would also explain the inverse-square laws, tracing them to the inverse squares in Bohr's formula and the time dilation that appears in the graphical account.

The Fine Structure Constant

The fine structure constant is a dimensionless number, 137. Richard Feynman called it "the greatest damn mystery in physics." It is necessary for the calculations of quantum electrodynamics, but it is not the count of anything in his Feynman diagrams. However, 137 is the count of arrows in the time diagram in figure 10.

On the left in figure 10 is a region of four-dimensional lattice consisting of seventy-eight arrows. Displaced to the right are the fifty-nine arrows that connect vertically aligned nodes, quanta in the proper-time axis. The sum is 137 arrows exhibiting the fine structure of one cycle of the cloud of a helium atom.

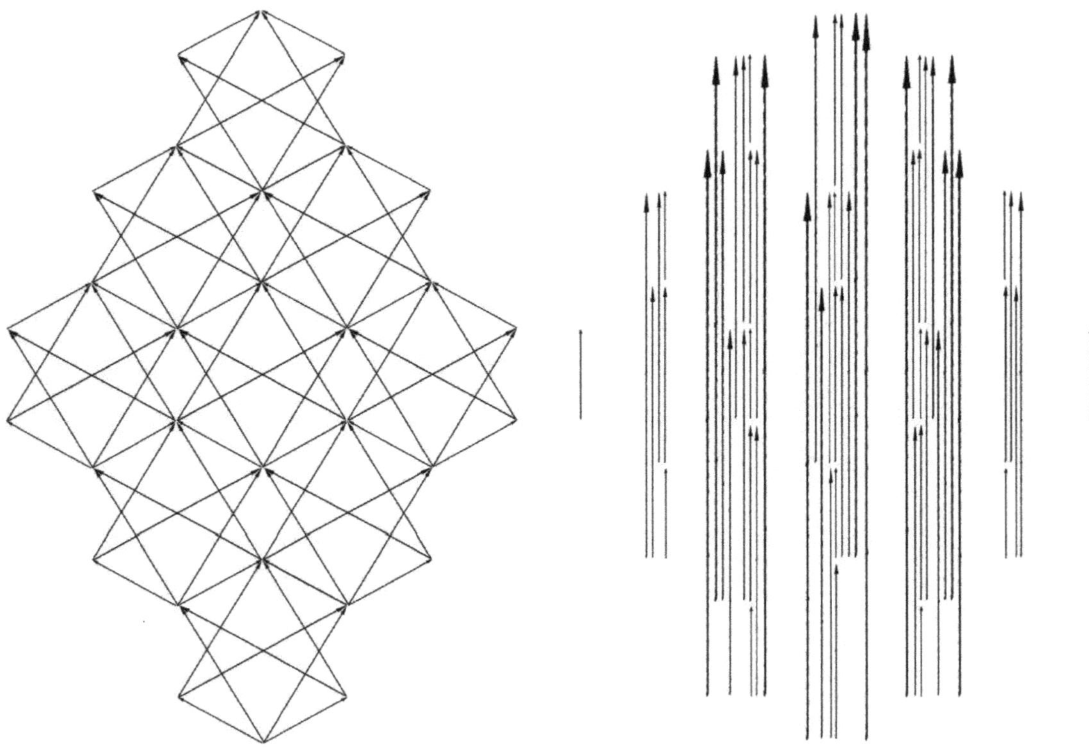

Figure 10. 137 Arrows.

The Nuclear Structure

Attached to periodic nodes of an electron cloud sequence is a nuclear sequence composed of higher-frequency quanta. This constitutes a synchronization of the nucleus to its electron cloud, assigning the nucleus its location in the four-dimensional manifold. The attachment of nucleus to cloud can only mean that the two distinguishable sequences share periodic nodes.

Summary of the Constructions

Causal set constructions are relativistic by default, due to the exclusion of primitive spatial relations. Built from a successor relation, the theory has the same consistency as whole-number arithmetic. The particle constructions describe self-organized patterns formed by temporal succession. Symmetry is the only organizing principle guiding the constructions, so symmetry assumes the role of the forces. Since symmetries reduce the complexity of a formation, the burden of explanation on physics is reduced to a minimum.

We are apt to think of the pace of human moments as "just right" for coherent experience. We are reluctant to grant such experience to subatomic occasions because of their nanosecond quickness. But there is no absolute measure of duration. A duration only has meaning in ratio comparison to other durations. The pace of experience is just right for the occasions of any sequence, regardless of its frequency ratio to other sequences.

Common-sense realism regarding the existence of the physical world is not jeopardized by the reduction to time. The electrons and photons, the rock, the brain, the planets, and the galaxies all have their quanta and their mass-energy. None of it has instantaneous spatial extension because the chaining of temporal transitions provides all the extensiveness there is. There is no such thing as a physical state of instantaneous organization. What exists all at once is only each individual moment. Since each moment is generic and primitive, there is no specification of its physical state. To refer to a "physical state" is to freeze out time and immerse oneself in the illusion of spatial extension.

Descartes made use of a piece of wax in his hand in order to bring his notion of physical existence into stark focus. The rock in my hand serves just as well as a piece of wax. My rock is made of quanta. The constituent quanta of the highest frequency connect to form the quark cycles. The quark cycles combine to form protons and neutrons, and these combine to form nuclei. The nuclei are interlaced with electron clouds to form complete atoms, which in turn combine to form molecules. The molecular patterns connect to form the rock. An arrow diagram of my rock would show how all its quanta are connected into a single elaborate sequence. The quanta are not undefined. Each quantum is an irresolvable step of time sequence. My rock is a propagating time sequence, made of temporal transitions from one moment to another. As it is with the rock, so it is with my hand that holds the rock, my body, physical objects in general, and the whole universe.

The key argument for panpsychism concerns the compatibility of human mental events with their immediate causes and effects. It has been a standing argument against mindlike entities that they are unlike the spatial entities of physics, so that causal interaction between the two is unintelligible. But now "the shoe is on the other foot." A time series of human moments is well suited to instantiate the causal order, as detailed in the final section. The remaining moments of physical theory have no specified attributes whatsoever. It is these latter moments that now stand in need of causal compatibility with mind. As for the causal pairing relations, since they connect mental event to mental event, they must be generally mental as well. To suppose otherwise would be to inject incompatibility for no reason.

Issues between Whitehead and the Reduction of Physics to Time

Hypothesis, speculation, conjecture: these roughly equivalent words express support for the essential role of hypothesis in theoretical physics.

Speculative Philosophy can be defined as the endeavor to frame a coherent, logical, necessary system of general ideas in terms of which every element of our experience can be interpreted. Thus speculative philosophy embodies the method of the "working hypothesis." (AI 22)

When Whitehead argues for a special sense of causation, "non-sensuous perception," he is attempting to justify belief in the causal derivation of our sensory data without resorting to hypothesis. "We see with our eyes." But intervening between the eyes and the phenomenal percept are the optic nerves, which we do not perceive. Whitehead champions speculation anyway. "There is nothing in the universe other than instances of passage and components of those instances." The reduction to time puts the full weight of physics behind that speculation. We now have the "coherent, logical, necessary system of general ideas in terms of which every element of our experience can be interpreted." The coherence of the scheme provides all that one could want to justify the belief that there are external causes to our phenomenal sensory data.

Whitehead retained four-dimensional extension as a characteristic of occasions. This fails to deal with Descartes's extension in space as the defining characteristic of "the physical." It maintains "dualism all the way down." The patterns of temporal succession vary in their dimensionality: the one-dimensionality of a light ray, the four-dimensionality of the manifold, and the six-dimensionality of the nuclear structure. Occasions, depicted by nodes in the arrow diagrams, anchor all the relational patterns, but it is not the occasions that have the dimensions.

Dimensionality is a structural feature of the patterns formed by temporal succession. If occasions do not have the physical attribute of extension, then there is no overlap of occasions, no "extensive abstraction." The next-to-next linking of occasions by causal inheritance relations, physical prehensions, is all that is needed to form the universe. Any of the arrow diagrams is an inheritance map, with each arrow depicting an inheritance passed from one occasion to a successor occasion. There is nothing physical to be passed from one occasion to another, because the theory of physics is already exhausted by the bare inheritance map itself. The form and content of subjective experience is all there is to an occasion, so the occasion has nothing else to give, nothing else to render to its successors.

Smooth Sailing

In the thirteenth century, the Buddhist philosopher Dogen formulated the idea of a causal web of mental events: "Mind is a knot in the net of interdependent origination." That is quite an accurate statement of Whitehead's theory of occasions. We now have the confirmation of this in physics. It is no accident that the fundamentals of physics can be derived from patterns of temporal succession.

Here is a summary of what we have so far, as I wrote in a previous work:

In the light of the foregoing, let us reconsider the notion of the physical world described in Chapter 2. I said there that the meaning of "physical" amounts to an intuition of space and what is in it. Special Relativity implies that determinate location in space-time is due to time-ordering relations alone, as indicated by the arrow diagrams. The theory abandons purely spatial relations that define space as instantaneous extension. Since one event can have several immediate causal predecessors or successors, we must conceive the course of time as branched into locally separated streams. For physics, spacelike structure is part and parcel of this richer structure of time. The notion of space as an extended state is abandoned. This demands a corresponding revision to our conception of physical entities as inherently spatial. That conception belongs to a provisional stage of science that has been overturned by the discovery of a limiting velocity.

The analysis of the universe into whole and part now finds the parts to be immaterial events that occur in somewhat regular patterns of succession. These patterns account for physical space, energy, shape, location, and motion. This constitutes the dematerialization of matter and the dismantling of physical space required by Special Relativity. The meaning of "physical" therefore devolves upon events, their time relations, and the resulting patterns. With the understanding that causal succession is equivalent to temporal succession, physics amounts to a theory of what comes before what. Progress in this theory has arrived at quantum events, discrete events which do not admit of further before-and-after analysis. These rudimentary events, and the time relation that orders them, are the quintessential physical entities. Our impression of the physical world as something substantial and immense owes to the sheer number of quantum events and their causal connections. If the subject of scientific investigation is simply the temporal sequence of events, then the physical world is not well characterized as "space and what is in it" but rather "time and what is in it." This has immediate consequence for the mind-body problem, since there is little difficulty in assigning the time of occurrence to a mental event. (Carlson, Mind-Body 73–75)

The Frequency and Duration of Human Occasions

Whitehead died before the cortical homunculi were discovered, and he never knew that temporal succession forms frequency ratios. By giving the discrete event ontology "the widest extension of which it seems capable," we can treat the brain as a purely sequential formation and describe in detail the causal location and functional role of human mental events in the brain (see Carlson, "Finite").

If one person's sentient mind is to be considered a discrete temporal series, the first order of business is to determine the frequency of the human series with respect to the other frequencies of physics. Psychophysical experiment indicates that 10 Hz is the frequency of human occasions. Let us consider the experimental data.

The closest physical correlate to human sentient awareness is the brain-wave activity at the cortical surface, recorded by an electroencephalograph (EEG). In the late stages of amyotrophic lateral sclerosis (ALS), a patient can reach a locked-in stage, having lost the last vestige of voluntary motor control. The patient's mind is stranded, without even an eye blink to communicate to the outside world. In such a case, the brain-wave activity indicated on an EEG is the only means by which a doctor can determine whether the patient's brain is still host to a sentient human mind.

Dreamless sleep is indicated by slow rolling waves of several cycles per second. Alpha frequencies of 10 Hz accompany both the waking condition and active dreaming. The alpha frequencies are typically out of phase with one another, making a jumbled mess on the EEG record. A good meditator can bring the alpha frequencies into synchrony, producing a steady rhythm of coherent oscillation at 10 Hz. That is one clue to the frequency of human occasions.

A stimulus probe on the visual cortex produces a spot in the subject's visual field. A pulsing stimulus repeated at one location produces a pulsing spot for the subject until the pulse rate exceeds ten per second. As that stimulus frequency is approached, the sensory spot loses its frequency altogether and becomes steady. Testing the frequency response of the other sensory modalities reveals the same 10 Hz limitation. The implication is that we cannot register changes that are quicker than ten per second. The likely explanation: moments of human sentient awareness transpire at that rate and no faster.

A rate of ten per second for human moments is also appropriate to the delays involved in the conduction of efferent nerve signals from the brain to the muscles, and in the reverse direction, afferent signals from sense receptors to the brain. Reaction time, to avert a driving collision, for example, is not reducible to less than one tenth of a second. Reliable motor control of the body requires patience for the feedback, which is subject to the propagation delays of neural transmission. The human series is well qualified for central control of the human body, equipped at 10 Hz as the ideal frequency for the job.

Strobe lights at 10 Hz bother people, and epileptics are prone to seizure when they see such lights. All in all, given that we are seeking a finite frequency for the human series, a regular frequency of 10 Hz seems to be it. We are not aware of this frequency by introspection. It is ascertained only by reference to scientific hypotheses concerning a world that lies beyond the reach of anyone's introspective powers.

Location of the Human Series in the Brain Sequence

Brain scientists have mapped out a set of functional locations on the cortex called projection areas. These serve to pinpoint the location of the human series in the brain. The first two projection areas to consider are depicted by the motor homunculus and the sensory homunculus (see Figure 11), which represent humanlike forms on the cortex that were first mapped out by Wilder Penfield. The topology of the human body is preserved in these shapes, but geometric distortions of the "little man" give him the appearance of a malformed fetus.

You can stimulate the motor homunculus with a probe and get the corresponding part of the body to twitch into action, like operating a puppet. You can stimulate the sensory homunculus to shortcut the more remote stimulus that is normally needed on the surface of the body to achieve the same sensation. Each moment of a human series has additional predecessors and successors that belong to the brain, but not to the human series. Forking and convergence connect the human series to other cycles of the brain.

Quanta that fork off from the human series to the motor homunculus affect bodily movement. Quanta from the sensory homunculus converge upon the human series, updating the body image of somatic awareness.

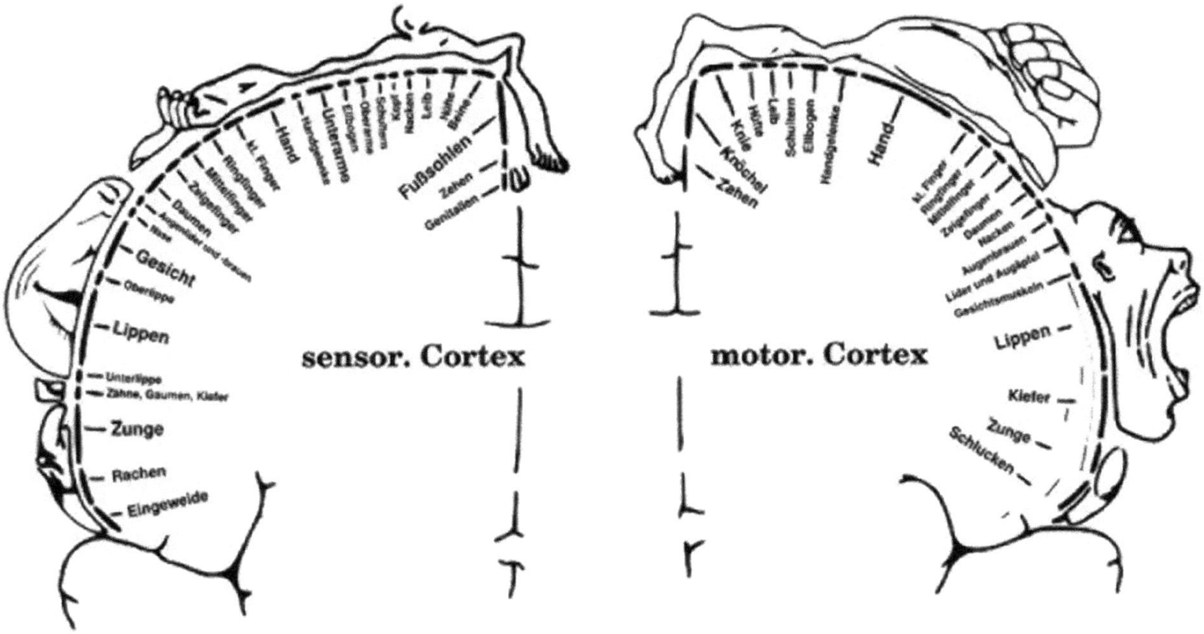

Figure 11. The Cortical Homunculi.

At cycles of 10 Hz, the sequence of cause and effect is as follows:

- One human moment forks off, connecting to many moments of the motor homunculus.
- Effects are propagated along efferent nerve routes to the muscles.
- Muscle action causes feedback signals along afferent nerve routes to the sensory homunculus.
- Many moments of the sensory homunculus converge upon the next moment of the human series.

During the one tenth of a second between the two bounding moments of the above cycle, one human quantum also transpires, propagating the human series. The spacelike equivalent of a tenth of a second is about 18,000 miles, plenty of range to gather concurrent influences from the surface of a cortex.

The standard conception of a brain is one of instantaneous extension in space, with no earlier and later involved in its composition. That is a brain without quanta. Such a brain has no place in our physics. According to Special Relativity, the cortical surface is a set of contemporaries, causal cousins related only by their causal ancestry. Such contemporaries are also poised to beget common causal descendants.

The location of the mind in the brain is resolved by tracing the causal lineage of human mental events to and from the homuncular regions. The homunculi on the cortex are the key causal locators of human mental events. Other projection areas on the cortex have also been mapped out that correspond to other sensory fields of human phenomenology. Patterns of excitation at the retina are reproduced at the visual projection area. Auditory experience also has a patch of cortical surface devoted to it. A mental event typically involves all the sensory modes at once.

The distinct phenomenal sensory modes correspond to the distinct patches of cortex devoted to the organs of sight, sound, and touch. As with the sensory homunculus, the visual and auditory projection areas are home to causal predecessors of the human series. From those cortical sites, the afferent system converges to a human percept, at which point the efferent phase of causal sequence is renewed.

Let us consider visual experience and its patch of cortex. In phenomenal vision, we have a spatially extended field of colored patches. The colorful visual field is part of a human mental event. As with color, we are directly acquainted with the inherent geometry of the visual field. We can judge with remarkable precision the size and shape of colored patches given in our visual experience.

A good example is the extraordinary precision by which we can judge a rectangle to have height-width proportions of the golden mean. The ancients could judge with the same accuracy. It owes nothing to science. It is a type of ratio measurement that involves no scientific conceptions or assumptions.

The visual projection area has a space-time metric based on the second as the standard unit of duration. The speed of light is one foot per nanosecond. Supposing the patch of visual cortex to be roughly circular and one inch in diameter, its spacelike extent is approximately one tenth of a nanosecond. The full spread of the subject's phenomenal visual field correlates to the full diameter of the cortical patch, so half the visual field corresponds to half the cortical patch, or one twentieth of a nanosecond.

Proportionate size in the visual field is thus correlated to the metric unit of physics. This correlation is needed for an epistemological account of physical measurement, which requires sentient mental events in the laboratory and sensory data that are phenomenally given to them. The correlation of phenomenal measure to the nanosecond span of this or that cortical projection area is reliant on psychophysical experiments.

Perceivable sensory fields are correlated to the unperceivable domain of physics. We do not expand the domain of the perceivable by arriving at such correlations. In the case of hearing, a spacelike span across the auditory cortex correlates to an interval of phenomenal pitch. Psychophysics correlates the qualitative data of subjective experience to the conjectural model of theoretical physics.

The human series has direct access to vision, hearing, and tactile information at the sensory projection areas. "Direct access" is unambiguous in our theory of physics. It means that select moments of the sensory projection areas are immediate causal predecessors of a human moment. Each such predecessor connects to the human moment by a single quantum.

Likewise, direct action by a human moment upon some moment in the region of the motor homunculus means that a single quantum connects the human moment to the homuncular moment. The homunculi are situated on the cortical surface as if to provide convenient test points for a technician to troubleshoot the sensory and motor systems. In normal operation, they serve as staging areas for perception and control by the human series.

The stable brain is a propagation of synchronized time cycles, featuring a wide range of frequencies and likely a wide variety of cycle topologies. The stability of human experience and its dependence on the brain means that the human series must be embedded in supportive cycles of 10 Hz frequency. These cycles provide a base of causal routine for the human series and connect it to the ladder of higher frequencies that constitute the rest of the brain.

THE PRIMORDIAL NATURE OF TODAY'S FUNDAMENTAL PARTICLES

The symmetry and simplicity revealed in the causal set construction of the elementary particles have a corollary. The particles that dominate any stage or region of the universe must have the same structure as the particles dominant today. Photons, neutrinos, electrons, protons, and neutrons are each self-organized by chained repetition and time-reversal symmetry, representing the simplest patterns of discrete order that are mathematically possible. The structural building block for the particles is the quantum, the arrow of the arrow diagrams. The only organizing principle is symmetry, which assumes the role of the forces and laws of physics.

Causal Set Theory and the Origin of Mass-ratio

Quantum theory is reconstructed using standalone causal sets. The frequency ratios inherent in causal sets are used to define energy ratios, implicating the causal link as the quantum of action. Space-time and its particle-like sequences are then constructed from causal links, from quanta. A four-dimensional time-lattice structure is defined and then used to model neutrinos and electron clouds, which together constitute a four-dimensional manifold. A six-dimensional time-lattice is used to model the nucleons. The integration of the nucleus with its electron cloud affords calculation of the mass ratio of the proton (or the neutron) with respect to the electron. Arrow diagrams, along with several ball-and-stick models, are used to streamline the presentation.

The basis for the constructions is the frequency ratios formed by "the arrows of time" (causal set graphs). Time order and causal order are conflated in causal set theory.

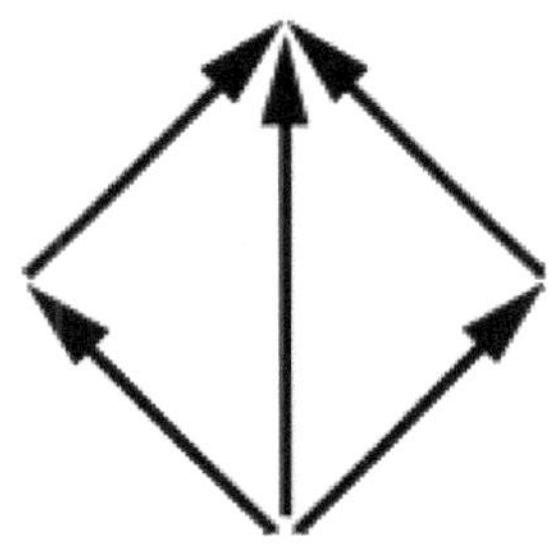

There are three causal routes from the bottom event to the top one. Either side route involves two causal transitions with an intermediate event along the way. All three routes determine the same temporal interval, since they all begin at the same time and end at the same time. Either side route has twice the temporal frequency of the middle route.

The next diagram of alternate routes between common end points has two arrows forming one route and three arrows forming another, producing a frequency ratio of 2:3.

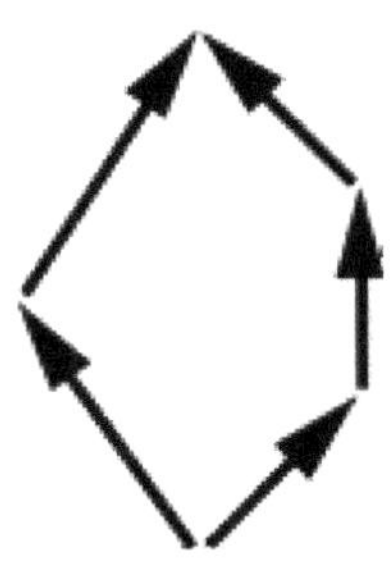

Time diagrams exist for frequency ratios ranging across the rational numbers. Reciprocals of the frequency ratios are wavelength ratios. "Wavelength" is a measure of duration in the causal set formulation of physics.

Planck discovered the coupling of energy and frequency at the quantum level, expressed in the formula E = hf. Thus, two energy values, E_1 and E_2, are equal to two frequency values, hf_1 and hf_2. The ratio of two energy values, E_1 / E_2, is equal to the ratio of two frequency values, f_1 / f_2. Planck's constant drops out. The frequency ratios can serve to define the energy ratios. Energy is just frequency. The step of time is the unit of the frequency ratios and the quantum of the energy ratios. We have a structural definition of energy and its quantum in terms of time alone. Directed graphs can now serve as quantum schematics, each arrow depicting a quantum.

Causal set constructions are relativistic by default. As Rafael Sorkin, creator of causal set theory, has said, causal set theory satisfies Special Relativity "out of the box." That is due to the exclusion of primitive spatial relations from causal set theory. The reduction to causal sets explains the limiting velocity, since any change of spacelike separation is a purely structural consequence of the stepping action of time (space cannot outrun time). With the causal link identified as the quantum of energy ratios, we also have quantum theory "out of the box."

It would be helpful to the physicists working on cold fusion to know the quantum structure of the nuclei. But the physicists are not cognizant of that structure. As one physicist said, "We don't know what particles are, we don't know what mass is, and we don't know why particles have the mass they do." A large part of their problem is the effort to obtain the orderly structure of atoms and molecules from the randomness and chaos thought to be endemic to quantum theory. Trying to obtain simplicity from complexity, or order from randomness, is wrong-headed. As a result, the working foundation of physics today is very shaky.

With the shaky foundation, the primordial nature of today's particles is not a given. It is said that the laws of physics break down inside a black hole. And physics is thought to break down at the singularity of the Big Bang. With the reduction of physics to sheer temporal/causal succession, those singularities seem to be unnecessary.

> Upon further reflection, it appears that the Big Bang has been misconstrued as a singularity marking the beginning of time, when it is more likely that our system of temporal succession has no beginning. As we consider earlier and earlier stages of the universe, we conceive a spatially shrinking universe, converging quite naturally according to our spatial intuitions to a point-like minimum of spatial extent. But to clarify the situation, the "shrinking of space" is a purely structural consequence of increasingly higher frequencies.
>
> A recent issue of Discover magazine featured a marble-sized sphere on the cover graphic that represents the spatial size of the universe at $t = 10^{-34}$ seconds. Yes, space shrinks toward zero as all frequencies increase, but clarity on the situation demands that we relinquish unanalyzed spatial intuitions. The negative exponent in the equation is heading toward larger integers (higher frequencies) as we delve into earlier stages of the universe.
>
> Nothing in the basis of our theory suggests that frequencies should have an upper limit unless we restrict consideration to a bounded region. Nothing suggests that the exponent in the equation will, with further knowledge on our part, attain some satisfactorily high integer and reach a "glass ceiling." Physicists of the distant future, should they be likewise captive to their spatial intuitions,

> might well conclude that our age was one of unbearable heat, a marble-sized universe suffering from excruciating proximity to a cataclysmic Big Bang. (A Theory of Everything for Physics, 2005)
>
> We think of the early universe, near the Big Bang, as cramped and hot, but time is stepping at a higher rate proportionately for all processes at earlier stages of the universe. The earlier frequencies are strictly relative to later frequencies because frequency arises solely in the form of frequency ratios. If all frequencies are proportionately greater in the distant past compared to today, what difference does it make to the production of fundamental particles? Everything was much the same then as now. It is as though we view the past through the wrong end of a telescope without being aware of it.

"Time flows into a black hole." That is one of my favorite statements from a physicist. What should we think goes on inside a black hole to give it mass-energy? Now that we have characterized simple quantum behavior outside a black hole, there is no reason to think that quanta form particles any differently inside a black hole.

The reduction of space-time to time alone is an idea that owes to the event ontology developed by Russell and Whitehead. A causal set graph of an extendable four-dimensional time lattice can help to overcome the conceptual hurdle of reducing physics to time.

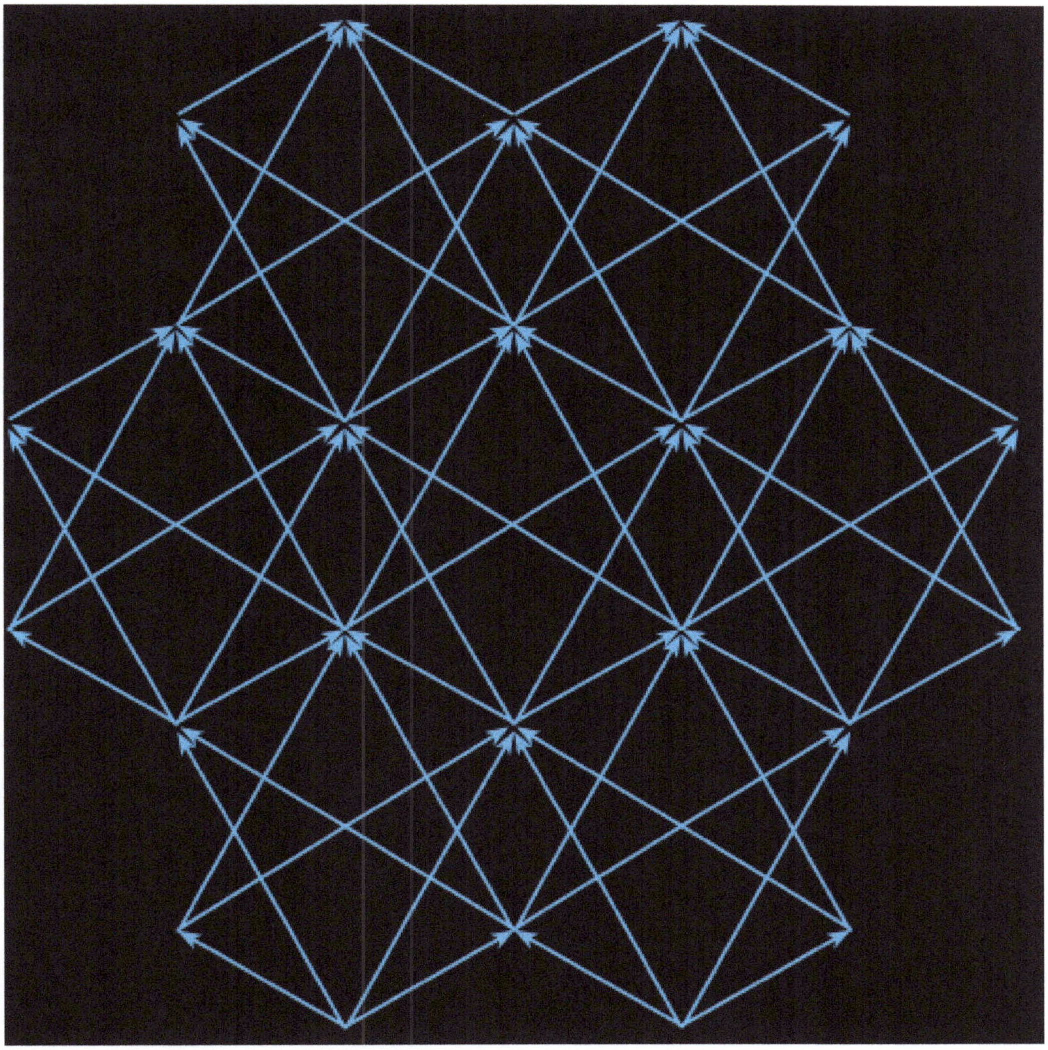

Every interior node is at the intersection of four axes of time. There are no primitive spatial relations from which spatial axes, or spatial entities of any kind, could be formed. The visible diagram is spatial, but purely phenomenal. The referent of the diagram is an invisible region of four-dimensional time, one of the theoretical entities of physics.

Carey R. Carlson, June 23, 2022

THE ROLE OF THE CAUSAL SET PROTON IN THE FORMATION OF 3-D MATTER

The reference frame of the earth gives us the impression of three-dimensional space and one-dimensional time. I propose to explain this in terms of the organizing role of the causal set proton in the formation of three-dimensional matter.

At the quantum level, there are no primitive spatial relations. Thus, there is no instantaneous three-dimensional spatial extension, at the quantum level or at any level. With the reduction to causal sets, the universe is simply a discrete causal web formed by the ongoing stepping action of time.

Subassemblies of the causal web can be graphed by "the arrows of time." Such graphs, using symmetry as the guide, have produced quantum schematics of the four-dimensional time manifold, the photon, neutrino, electron, neutron, and proton.

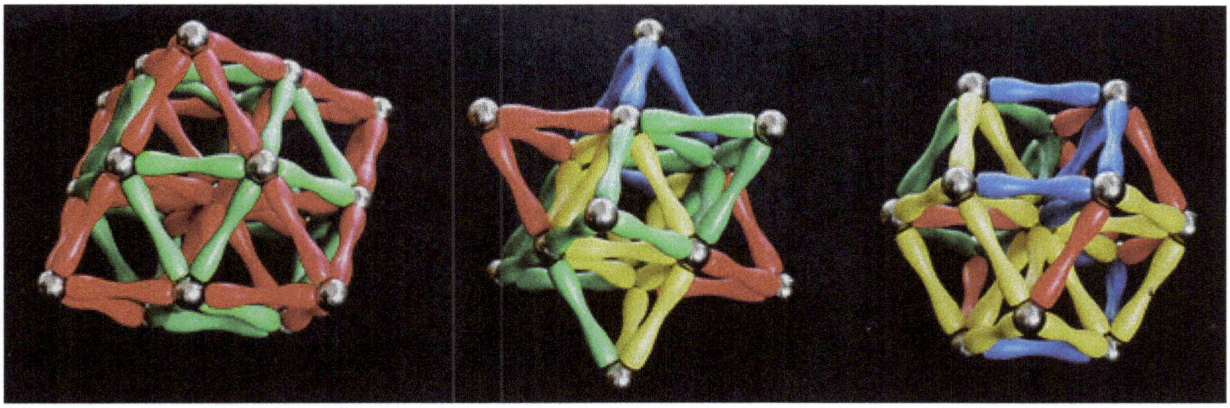

The proton is my focus here, the model in the middle. All three models are subassemblies of an extendable pattern called the octet truss. With the octet truss extended indefinitely, each subassembly will recur repeatedly, as will formations at twice the size, three times the size, and so on.

For any of the models pictured, you can convert the spatial wireframe to a causal set by picking any vertex and drawing a major time axis to the opposite vertex. Then you convert the rest of the struts to arrows, preserving as much symmetry as possible. In the case of the cuboctahedron on the right, you produce a perfect six-dimensional time lattice, extendable as you please, with every interior point at the intersection of six axes of time.

For the other two models, you would be choosing a different direction for the time axis in converting the octet truss to causal sets. The octahedron model, tripled in size, portrays a neutron. The model in the middle, compounded to a 3 × 3 × 3 larger cube, portrays a proton. That enlargement perfects the symmetry of the causal set versions and obtains the proper mass values for the neutron and proton.

Because the Periodic Table has the proton as its "index," the proper time axis of the proton likely sets the alignment of proper time sequences of the nuclei, the molecules, and solid matter, synchronizing the component processes. Solid matter needs an alignment of the proper time axes of its components, or it would fly apart, with pieces going off in different directions.

The six-dimensional proton has three-way symmetries of quanta about its proper time axis. The proper time axis has a neutral orientation with respect to three other mutually orthogonal axes. A massively parallel alignment of protons would effectively constitute a coherent global proper time (one-dimensional time) for the whole solid body.

The other three major axes of the proton, orthogonal to one another, provide the basic building block for the three-dimensional structure of the assembled solid.

If the time sequence of a proton is too quick to discriminate, we are liable to theorize that the proton is a three-dimensional spatial object of instantaneous mass persisting continuously through time. That is the Newtonian conception of a proton, which carries over to the Newtonian conception of the earth. It is better to construct everything from quanta, with all constructions relativistic by default.

Carey R. Carlson, January 4, 2023

DISCRETE TIME DILATION FROM CAUSAL SETS AND ITS ROLE IN DISCRETE GR

The recognition that frequency ratios are formed by the arrows of time led to the graphical account of Bohr's formula below. The "hex cycles," from which the sequences are compounded, are treated as electron cycles (although the charge quanta are not shown).

Each successive sequence has been scaled such that 36 electron cycles extend from the bottom of the page to the top, indicating that the separable sequences transpire in the same amount of time. Thus, taking the electron cycle as the unit of energy, an equal amount of energy plays out in an equal amount of time for any sequence. Discrete time dilation from one electron cloud to the next is represented.

My purpose here is to focus attention on neutrino formations rather than electron clouds. We will forget about Bohr's formula and the incident photon. The sequence on the right suggests that neutrino propagation can modulate freely between different forms. I will propose a causal set manifold composed only of neutrinos and electron clouds.

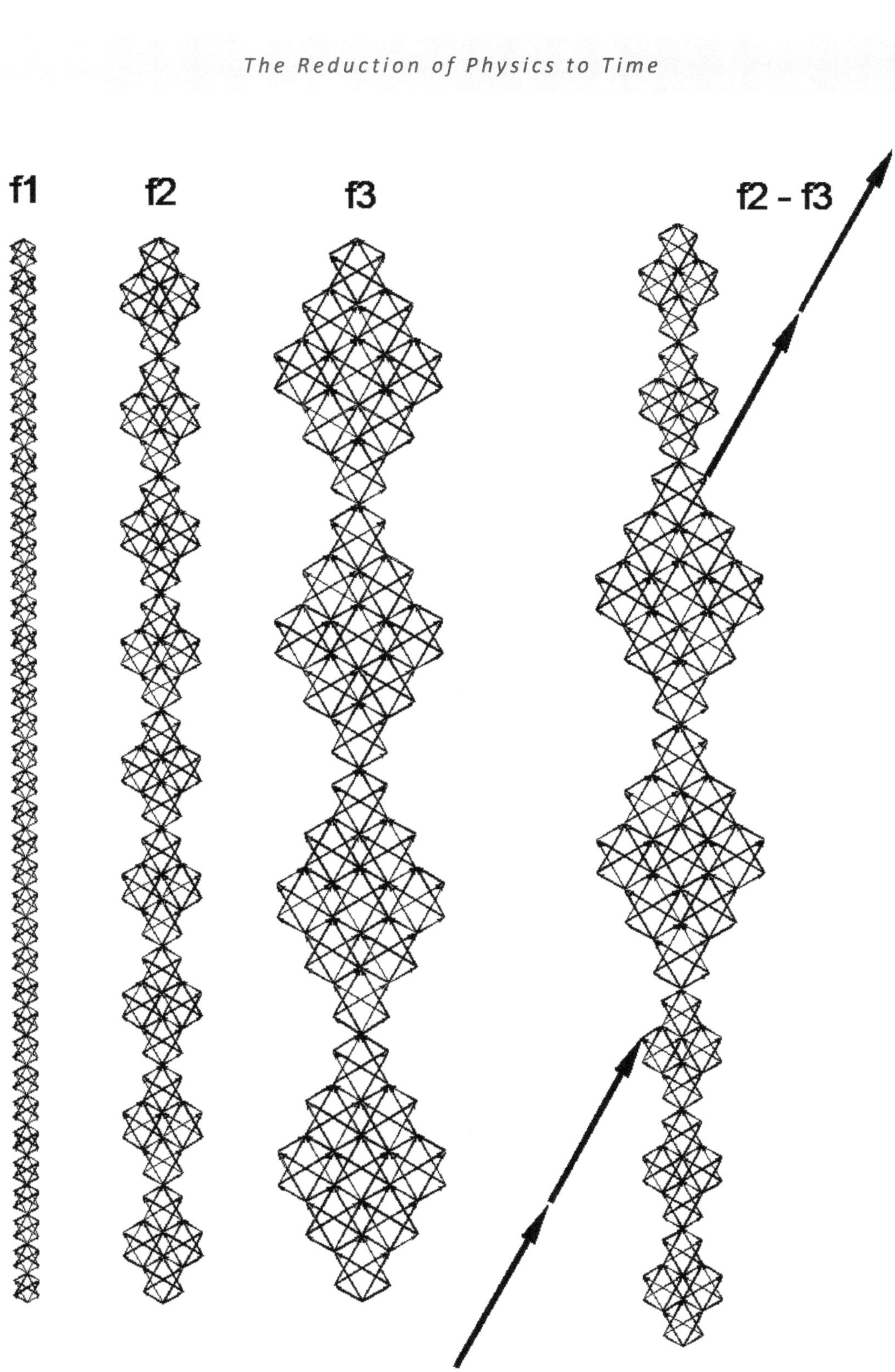
f1
f2
f3
f2 - f3

A region of uniform four-dimensional time is composed of neutrino cycles, as shown by this graph of an extendable four-dimensional time lattice.

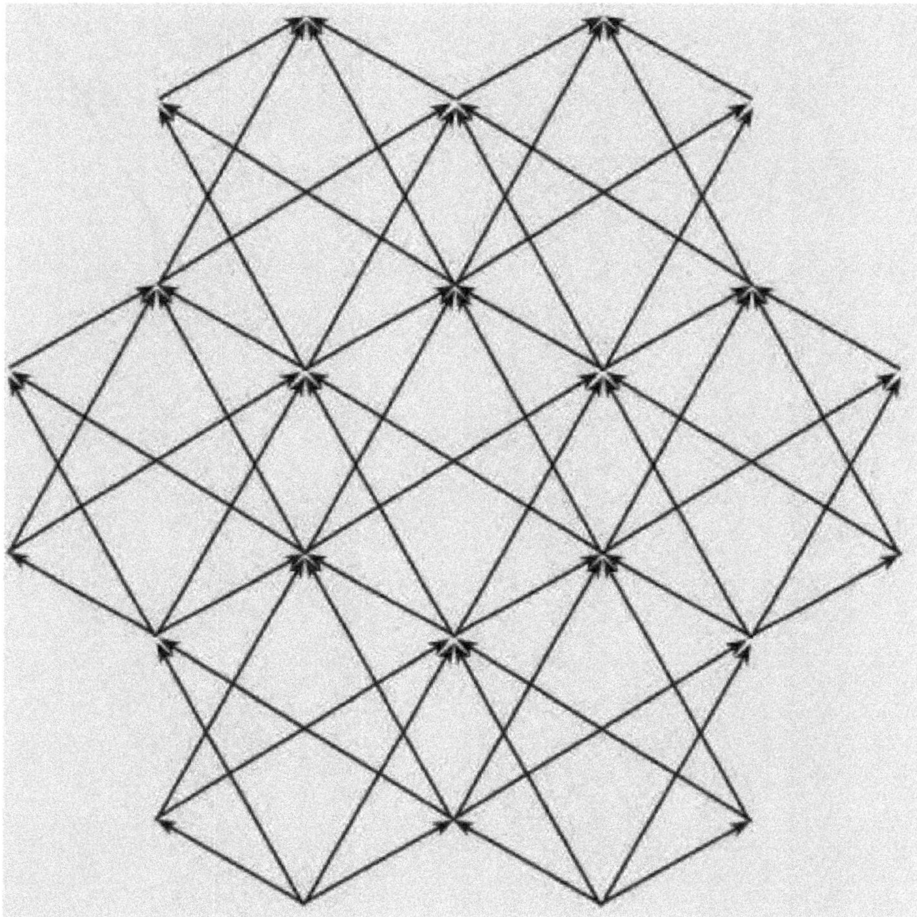

The presence of nuclear matter in the causal web disrupts the uniformity of the **four-dimensional** lattice due to time dilation. Nuclei dilate in step with their electron clouds, tying time dilation to mass in accord with General Relativity.

The uniformity of a region of four-dimensional time lattice is not geometrical, it is strictly topological. There is no background against which the discrete lattice can "stretch." The lattice must break apart, or split apart, for there to be any change of frequency, for there to be any time dilation of its cycles.

That which splits off is a form of neutrino propagation with its own frequency. Successive splitting progressively increments the time dilation until the higher-frequency lattice of a massless region meets up with, and matches, the lower-frequency lattice in the vicinity of a massive body.

Einstein's General Relativity employs a continuous manifold with continuous curvature to explain gravity. Using only fully discrete causal sets, the manifold must splinter into neutrino propagations to obtain a discrete analog of Einstein's General Relativity.

Carey R. Carlson, June 2, 2024

WORKS CITED

1. Carlson, Carey R. "Finite Eventism." Mind That Abides: Panpsychism in the New Millennium, edited by David Skrbina, Amsterdam, John Benjamins, 2009.

2. Carlson, Carey R. The Mind-Body Problem and Its Solution.Second edition, GoToPublish, 2019.

3. Russell, Bertrand. Human Knowledge: Its Scope and Limits. 1948. New York, Simon and Schuster, 1967.

4. Whitehead, Alfred North. Adventures of Ideas. 1933. New York, Free Press, 1967.

www.ingramcontent.com/pod-product-compliance
Lightning Source LLC
LaVergne TN
LVHW070855210926
848482LV00016B/945